NANOCOMPOSITES

NANOCOMPOSITES

In Situ Synthesis of Polymer-Embedded Nanostructures

EDITED BY

Luigi Nicolais
Gianfranco Carotenuto

WILEY

Published by John Wiley & Sons, Inc., Hoboken, New Jersey.
Published simultaneously in Canada.

For general information on our other products and services or for technical support, please contact our
Customer Care Department within the United States at (800) 762–2974, outside the United States at
(317) 572–3993 or fax (317) 572–4002.

Wiley also publishes its books in a variety of electronic formats. Some content that appears in print may
not be available in electronic formats. For more information about Wiley products, visit our web site at
www.wiley.com.

Library of Congress Cataloging-in-Publication Data:

Nicolais, Luigi.
Nanocomposites : *in situ* synthesis of polymer-embedded nanostructures/Luigi Nicolais,
Gianfranco Carotenuto. – First edition.
    pages cm
  Includes index.
  ISBN 978-0-470-10952-6 (hardback)
1. Nanostructured materials.   2. Polymeric composites.   I. Carotenuto, Gianfranco.   II. Title.
  TA418.9.N35N53 2014
  620.1′18–dc23
                                 2013016297

Printed in the United States of America.

10 9 8 7 6 5 4 3 2 1

CONTENTS

PREFACE

New physical and chemical properties arise in solid matter when it is reduced to a nanoscopic scale. Most of these novel characteristics (e.g., electron confinement, size-dependent magnetism, enhanced electric field near the surface, surface plasmon resonance) can be exploited for advanced functional applications in many technological fields. All material types (ceramic, polymers, metals, etc.) acquire unique properties at a nanometric scale; however, the most dramatic change in properties as a consequence of size reduction is evidenced by metals and semiconductors. Consequently, research activity in nanostructured materials worldwide focuses mainly on metallic and semiconductive nanoparticles. Because of their very small size, metal and semiconductor nanostructures are difficult to handle, store, and use. Single nanostructures can be manipulated by special techniques like surface tunneling microscopy (STM), dielectrophoresis, and magnetophoresis. However, even the fabrication of quite simple macroscopic devices (e.g. single-electron transistors) by these approaches results really difficult. The spontaneous arrangement of perfectly identical nanostructures in regular arrays (self-organization) is an important tool for the realization of devices based on nanosized objects. However, only a few spherically shaped nanosized solids are able to self-organize in two- or three-dimensional macroscopic structures. Nanostructures can be readily used if they are supported on special ceramic substrates or embedded inside organic or inorganic matrices (the embedding phase can also be a liquid, as in the case of ferrofluids). Supported nanostructures are used mainly in the fabrication of heterogeneous catalysts and special gas-absorbing media, but in all other cases the preferred approach to handle nanoparticles is to have them embedded.

In addition polymer-embedded nanostructures are chemically protected and stabilized. Practically, all atoms and molecules in nanosized solids are located at the surface. Thus, they are coordinatively unsaturated and therefore extremely reactive. For example, oxidation and contamination by absorption of small molecules present in air (e.g., SO_2) is a very fast process for metallic nanopowders. In addition, reactions involving nanosized solids stoichiometrically behave and also noble metals like gold, palladium, and platinum are not anymore chemically inert on a nanometric scale. Embedding of nanoparticles is a very simple way to solve all reactivity problems.

Toxicity is another important aspect concerning nanostructures, and the embedding process also presents a valid solution to this problem. This is an emerging topic in the field of nanotechnology and will determine future developments in this research area. Because of toxicity, the handling of large amounts of nanostructures will be very critical and their use in an embedded form represents a really promising approach. The risk of using toxic nanostructures is expressed as the product of hazard and exposure ($R = H \times E$).

The embedding process reduces exposure, thus decreasing the risk drastically. In the future, embedding of nanostructures will probably be the only safe method for using nanoscopic materials.

So far, research activity on embedded metal and ceramic nanostructures has mostly been done by inorganic hosts (typically glasses). Inorganic glasses adequately protect nanoparticles, but they have two main disadvantages: (i) processing of these nanostructured materials is quite complex for the high temperatures involved, and (ii) glasses have scarcely tunable physical and mechanical characteristics. Polymer embedding of nanostructures shows several advantages in comparison with glass embedding. As polymer embedding requires only moderate temperatures for processing and has wide-ranging physical characteristics, they allow the nanocomposites to be suitable for many technological applications. For example, polymers can be dielectric media or dielectric conductors, hydrophobic or hydrophilic solids, hard–fragile (thermosetting resins), soft-plastic (thermoplastics), or elastic (rubbers) materials. Such a variety of properties is of fundamental importance to design functional materials. In addition, polymers are permeable to small molecules, which presents great potential in the fabrication of chemical sensors for gasses and liquids.

Unlike their chemical properties, a number of physical properties of nanostructures remain unmodified after polymer embedding (e.g., magnetic, optical, dielectric). In addition, owing to the large interface area, there are several polymer characteristics that are affected by filling with nanosized solids (e.g., glass transition temperature, crystallinity content, flame resistance, and gas permeability).

Nanostructures have tremendous potential for functional applications. However, the use of these materials is still very limited due to the difficulty in tuning their properties by controlling their morphological/structural features. In addition, since the collective properties of nanosized systems strongly differ from those of single particles, the special topology realized in the nanocomposite material represents a further way of modulating functional characteristics. Such a unique composite class represents a revolutionary approach in the science and technology of materials, and it has enormous potential for advanced functional applications. In fact, polymer-embedded nanoscopic structures like metal clusters, semiconductors, quantum dots, fullerenes, fine granular superconductors, ceramics, nanoparticles, nanotubes, nanoshells, graphenes, etc. can be exploited for functional applications in a variety of technological fields (e.g., optics, electronics, and photonics). However, all aspects concerning functionality control, multifunctionality, etc. are known only at a pioneering level, and systematic investigations are required.

The functional properties of polymer-based nanocomposites arise from the nanoscopic filler or from a combination of polymer and filler characteristics. Different approaches are available for the preparation of nanostructures; however, chemical methods are most effective for the large-scale production of these special solid structures. In fact, chemical methods allow bulk quantities of reproducible material to be obtained. Since functionality control is extremely important and the properties of solid phases at a nanoscopic scale are strictly dependent on size and shape, it is imperative to develop preparative schemes that allow one to have control over the product morphology.

Nanocomposites are potentially useful for many advanced functional applications in different technological fields because of (i) the optical transparency related to the small filler size, (ii) the unique nanophase functionality due to surface and confinement effects, and (iii) the variety of properties that organic–inorganic hybrids may have. A description for the main application areas follows.

Owing to large surface development, metal clusters have a special property known as surface plasmon absorption. Electron plasma oscillates at metal surface under the action of electric field of light, and resonances are possible at particular frequencies. Consequently, metal nanoparticles can absorb light by this special mechanism. Clusters of coin metals (gold, silver, copper) and some of their alloys (e.g., Pd/Ag and Au/Ag) strongly absorb light; therefore, they can be used as pigments to fabricate optical limiters (color filters, UV absorbers, etc.). The main advantages in comparison with traditional organic dyes are as follows: (i) they have very intensive coloration (e.g., the optical extinction of silver is c. 3×10^{11} $M^{-1}cm^{-1}$), (ii) high transparency, and (iii) light fastness and (iv) provide the possibility of making ultrathin colored films. The use of alloyed metal clusters (e.g., gold–silver and platinum–silver alloys) makes it possible to fine-tune the maximum absorption frequency, covering the full UV-visible spectral range.

When particles are uniaxially oriented inside the matrix, two different resonance frequencies are possible (longitudinal and transversal plasma oscillations). Such property can be exploited for optical polarizer fabrication. Frequently, the uniaxial orientation of filler is achieved simply by cold-drawing of raw nanocomposite material (e.g., polyethylene-based materials). This regular morphology produces polarization-dependent optical properties, which allows one to make optical filters that are able to change color by modifying the light polarization direction. A number of electro-optical devices are obtained by combining liquid crystal displays with these special color filters (e.g., multicolor single-pixel displays).

Optical sensors are very promising devices since their use does not require an electronic apparatus. Different types of optical sensors (e.g., chemosensors, pressure sensors, and thermochromic materials) can be based on polymer-embedded gold or silver clusters. In these nanocomposite materials, the surface plasmon resonance frequency of metal clusters is strictly related both to interparticle distance and the host medium refractive index (r.i.). Polymers may undergo significant structural changes because of external stimuli (e.g., pressure increase, temperature changes, and fluid absorption). These structural modifications may influence the surface plasmon resonance of guest metal, producing a visible color variation. Thermochromic materials are systems made of metal clusters topologically organized in an amorphous polymer in the form of extended planar aggregates. Metal clusters are not sintered together but are separated by a thin organic coating layer. Below the organic coating melting point, the material is characterized by a "collective" surface plasmon absorption of aggregated clusters, while above this temperature the expansion of the coating does not allow particles to interact with each other, leading to the coloration of "isolated" particle. Polymers may swell by absorption of fluids having comparable polarity. If these polymers incorporate Au or Ag clusters, their coloration changes during swelling owing to interparticle distance and r.i. modification. A variety of plasmon-based chemosensors can be made using polymers of different polarities. These materials are especially useful for clinical applications as disposable rapid test strips. Rubbers are amorphous polymers that crystallize under stress, and when filled by a great amount of coin metal clusters, change coloration depending on the stress applied. Such systems can be used as pressure sensors and devices to measure deformation. In general, polymer sensors are convenient due to their low price, disposability, easy fabrication, possibility to be applied by spry, etc.

Thermal reflectors are important devices that protect electronic equipment against exposure to intense solar light or fires, cryotechnological applications, solar energy exploitation, military uses (IR shields), etc. The possibility of fabricating these systems using polymers is of fundamental importance since polymers can be manufactured as textiles, applied as varnish, easily processed in a variety of shapes, and used as adhesives, sealants, etc. Gold-based nanocomposites with special topological features (i.e., presence of extended planar aggregates) show a strong ability to reflect near infrared (NIR) radiation. These materials can be easily produced on a large scale by thermal decomposition of special gold precursors. If optical plastics are used, the resulting nanocomposite material will serve as an effective infrared barrier (heat mirrors) but will be transparent to visible and UV light.

Metal clusters are so small that their nanocomposites are optical media. Nanosized metals are characterized by an r.i. value that ranges from less than 1 to ca. 6 in the visible spectral region. Metals like silver and gold have very low r.i. values ($n_{Ag} = 0.01$, $n_{Au} = 0.5$), and metals like tungsten and osmium have very high r.i. values ($n_W = 4$, $n_{Os} = 6$). At low filling factors (less than 15% by weight), the r.i. of polymer-embedded metal nanoparticles is given by the linear combination of pure bulk-metal and polymer r.i. values, but a significant deviation from this simple mixture rule has been observed at higher metal loadings. Since all optical plastics have r.i. close to 1.5, the possibility of modifying these values by introducing metal fillers is very important. Polymeric materials with ultra-high/infra-low r.i. values are very useful, for example, for waveguide fabrication (optical fibers require a high-r.i. material for core and a low-r.i. material for cladding). Such plastic fibers offer a number of advantages in comparison with traditional inorganic optical fibers: low-price, good mechanical performance, easily continuous production by coextrusion of molten metal/polymer blends, etc.

Ferromagnetic particles of iron, nickel, cobalt, gadolinium, chromium and manganese alloys, lanthanides, etc. reduced to a nanometric size do not produce scattering phenomena; therefore, they can be embedded into optical plastics to give magnetic materials transparent to visible light. Because of transparency, such magnetic plastics may have important magneto-optical applications such as Faraday's rotators. In particular, dielectric materials like polymers produce low-intensity Faraday's effect (i.e., when placed into an intensive magnetic field they rotate plane-polarized light, but the characteristic Verdet's constant value is low). The presence of a ferromagnetic filler inside these polymers significantly increases the Faraday effect, giving adequate nanocomposite materials for magneto-optical applications like optical windows (ultrafast shutters), optical modulators, optical isolators, etc.

Materials characterized by optical properties depending on the intensity of incident light are strongly required in different technological fields. For example, organic solar cells (e.g., ftalocianine-based devices) may be damaged by exposition to very intensive sunlight. The protection of these photovoltaic devices from intensive light exposure requires special optical limiters which absorption coefficient behaves non-linearly with light intensity. Polymer-embedded semiconductors, such as lead sulfides (PbS), are ideal material for such an application. Optical filters based on PbS are almost transparent under scarce illumination conditions and develop strongly absorbing properties in the presence of intensive sunlight. In addition to extinction, many other optical properties have been found to nonlinearly behave in optical plastics filled by little amounts of metals and semiconductors (e.g., r.i.).

Polymer-embedded semiconductor nanoparticles (quantum dots) like sulfides and other chalcogenides may emit monochromatic light in the visible spectral region when exposed to ultraviolet radiation. Since the band gap in these semiconductors is size-dependent, the color of emitted light can be tuned by controlling the dimension of the nanoparticle. These nanocomposite materials can be used as high monochromatic light sources for special photonic applications. By incorporating semiconductor nanocrystals in carrier-transporting polymers, an interesting class of photoconductive nanocomposites can be created. The presence of semiconductor nanocrystals enhances the photo-induced charge generation efficiency and extends the sensitivity range, while the polymer matrix is responsible for charge transport. A wide variety of semiconductors (e.g., CdS, GaAs, HgS, InAs, Ga_2S_3, and In_2S_3) and conductive polymers (e.g., polypyrrole, polyvinylcarbazole, polyaniline, polythiophene, and amine-doped polycarbonate) can be combined together for this application.

In conclusion, the physics and chemistry of nanometer-scale objects, and materials and devices that can be obtained from them, represents one of the most important areas of new material research. In particular, the science of mesoscopic metal structures has enjoyed an explosive growth in recent years due to the promise it has shown in major applications in electronics and photonics. Nanostructured metals are characterized by novel properties that are produced by surface and confinement effects. Many of these characteristics can be used to generate polymers with advanced properties. The resulting polymer-based nanocomposites are very promising as advanced functional materials in many technological fields. In addition, they have the processability of polymers, the outstanding electrical and optical properties of metals, and the lack of light scattering of nanometric fillers.

Luigi Nicolais
Gianfranco Carotenuto

CONTRIBUTORS

A. Alonso
Universitat Autònoma de Barcelona
Barcelona, Spain

D. Altamura
Institute of Crystallography
CNR, Bari, Italy

S. Badilescu
Optical Bio-Micro Systems Laboratory
Department of Mechanical Engineering
Concordia University
Montreal, QC
Canada

G. Carotenuto
Institute for Composite and Biomedical Materials
National Research Council
Napoli, Italy

W. R. Caseri
Department of Materials, ETH Zürich
Zürich, Switzerland

L. Cristino
Istituto di Cibernetica – CNR
Pozzuoli, Napoli, Italy

G.-L. Davies
Trinity College Dublin
Dublin, Ireland

S. De Nicola
CNR-SPIN Napoli
Complesso Universitario di Monte Sant'Angelo via Cinthia
Napoli, Italy

INFN-Sez. di Napoli
Complesso Universitario di Monte Sant'Angelo via Cinthia
Napoli, Italy

M.A. Di Grazia
Istituto di Cibernetica – CNR
Pozzuoli, Napoli, Italy

G. I. Dzhardimalieva
Institute of Problems of Chemical Physics
Russian Academy of Sciences
Chernogolovka, Moscow
Russia

C. Giannini
Institute of Crystallography
CNR, Bari, Italy

Y.K. Gun'ko
Trinity College Dublin
Dublin, Ireland

E. Hariprasad
School of Chemistry
University of Hyderabad
Hyderabad, India

T. Hasell
Department of Chemistry
University of Liverpool
Liverpool, UK

J. Macanás
Universitat Politècnica de Catalunya
Barcelona, Spain

M. Muñoz
Universitat Autònoma de Barcelona
Barcelona, Spain

D.N. Muraviev
Universitat Autònoma de Barcelona
Barcelona, Spain

F. Nicolais
Dipartimento di Scienze Politiche Sociali e della Comunicazione
Università degli Studi di Salerno
Fisciano, Italy

L. Nicolais
Dipartimento di Ingegneria dei Materiali e della Produzione
Università "Federico II" di Napoli
Napoli, Italy

M. Packirisamy
Optical Bio-Micro Systems Laboratory
Department of Mechanical Engineering
Concordia University
Montreal, QC
Canada

M. Palomba
Institute for Composite and Biomedical Materials
National Research Council
Napoli, Italy

A. D. Pomogailo
Laboratory of Metallopolymers
Institute of Problems of Chemical Physics
Russian Academy of Sciences
Chernogolovka, Moscow
Russia

A. Pucci
Macromolecular Science Group
Department of Chemistry and Industrial Chemistry
University of Pisa
Pisa, Italy

INSTM, Unità di Ricerca di Pisa
Pisa, Italy

CNR NANO, Istituto Nanoscienze-CNR
Pisa, Italy

T. P. Radhakrishnan
School of Chemistry
University of Hyderabad
Hyderabad, India

G. Ruggeri
Macromolecular Science Group
Department of Chemistry and Industrial Chemistry
University of Pisa
Pisa, Italy

INSTM, Unità di Ricerca di Pisa
Pisa, Italy

H. SadAbadi
Optical Bio-Micro Systems Laboratory
Department of Mechanical Engineering
Concordia University
Montreal, QC
Canada

A. Satti
Trinity College Dublin
Dublin, Ireland

D. Siliqi
Institute of Crystallography
CNR, Bari, Italy

R. Wüthrich
Optical Bio-Micro Systems Laboratory
Department of Mechanical Engineering
Concordia University
Montreal, QC
Canada

1

METAL–POLYMER NANOCOMPOSITES BY SUPERCRITICAL FLUID PROCESSING

T. Hasell

Department of Chemistry, University of Liverpool, Liverpool, UK

1.1 INTRODUCTION TO POLYMERS, NANOPARTICLES, AND SUPERCRITICAL FLUIDS

Nanoparticles
A nanoparticle is defined by the British Standards Institution as follows:
Nanoparticle—particle with one or more dimensions at the nanoscale
Nanoscale—having one or more dimensions of the order of 100 nm or less
This is in good agreement with how the term is used in general within the scientific community, although there is some degree of ambiguity as to the upper size limit. Particles and materials with the smallest domain sizes up to a micrometer and even several micrometre are sometimes referred to as "nano," although this is becoming less common with the increasing standardization of terminology in nanoscience.

1.2 PROPERTIES

Over the last three decades, nanoparticles have received an increasing amount of research interest. This is due to the unique size-dependent properties of nanoparticles, which are often thought of as a separate and intermediate state of matter lying between

Nanocomposites: In Situ *Synthesis of Polymer-Embedded Nanostructures*, First Edition.
Edited by Luigi Nicolais and Gianfranco Carotenuto.
© 2014 John Wiley & Sons, Inc. Published 2014 by John Wiley & Sons, Inc.

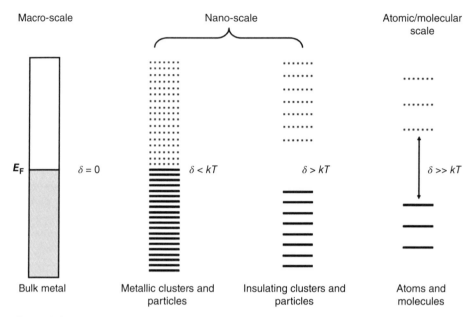

Figure 1.1. The evolution of the band gap and the density of states as the number of atoms in a system increases (from right to left). E_F is the Fermi energy level of the material and δ is the Kubo gap; see related text. Adapted from Roduner [2]. Reproduced by permission of The Royal Society of Chemistry.

individual atoms and bulk material. The properties of nanoparticles arise as a consequence of the confinement of the electron wave function and of the extremely high proportion of surface atoms—both of these factors are directly dependent on the size of the nanoparticle [1]. Indeed the possibility to control the properties, by tuning the size of the nanoparticle, has been the cause and subject of much investigation. Unlike bulk materials that have constant physical properties regardless of mass, nanoparticles offer unique opportunities for control by varying the diameter and have electronic, magnetic, and optical properties that can be manipulated. These effects occur because the energy levels for small particles are not continuous, as in bulk materials, but discrete, due to confinement of the electron wave function. The physical properties of nanoparticles are therefore determined by the size of the particle, the relatively small physical dimension in which the wave function is confined (Fig. 1.1).

The Fermi energy level (E_F) is the highest occupied energy level of the system in its ground state (lowest energy). The band gap (E_g) of these systems is the energy gap between the highest occupied and lowest unoccupied energy states. In these systems, from discrete atoms to bulk materials, the energy spacings are determined by the extent of the overlap between the electron orbitals of the material. Individual atoms have well-known atomic orbitals. These can combine in molecules to form molecular orbitals and further to form extended band structures, as in metals or semiconductors. The value of E_g is proportional to E_F divided by the number of delocalized electrons in the extended band structure. For a bulk metal, the number of delocalized electrons in the band structure is equal to the number of atoms in the bulk of the material. This normally results in E_g being very small and therefore only observable at low temperature. Under normal temperature, the delocalized electrons of the metal can be promoted easily to a higher energy state and can move freely through the structure. This gives the material its electrically conductive nature. In traditional semiconductor materials, the number of delocalized electrons is significantly less than the number of atoms. This results in a higher E_g that is significant at normal temperatures. This means that in a semiconductor the electrons will not be free to move, and conduct current, without some further input of energy. Equation 1.1 gives the average electronic energy level spacing of successive quantum levels (known as the Kubo gap):

$$\delta = \frac{4E_F}{3n} \qquad (1.1)$$

where δ is the Kubo gap, E_F is the Fermi energy level of the bulk material, and n is the total number of valence electrons in the nanoparticle

As an example, a silver nanoparticle of 3 nm diameter and ~1000 atoms (and therefore ~1000 valence electrons) would have a δ value of about 5–10 meV [1]. If the thermal energy, kT, is more than the Kubo gap, then the nanoparticle would be metallic in nature, but if kT fell below the Kubo gap, it would become nonmetallic. At room temperature, kT is ~26 meV, and therefore a 3 nm silver nanoparticle would exhibit metallic properties. However, if the size of the nanoparticle was decreased or the temperature was lowered, the nanoparticle would show nonmetallic behavior. Using this theory and a Fermi energy for bulk silver of 5.5 eV, then silver nanoparticles should cease to be metallic when under ~280 atoms, at room temperature. Because of the Kubo gap in nanoparticles, properties such as electrical conductance and magnetic susceptibility exhibit quantum size effects. These effects have led to nanoparticles being used for many applications from catalysis to optics and medicine.

1.3 CATALYSIS

The efficiency of materials already used as catalysts would be expected to be higher for nanoparticles than other solid substrates even by conventional theory. This is simply because nanoparticles have a much larger proportion of atoms in active surface sites compared to larger objects (see Fig. 1.2). Nanoparticles are so close in size to atomic dimensions that an unusually high fraction of the atoms are present on the surface. It is possible to estimate this fraction by using the simple relation shown in Equation 1.2:

$$P_s = 4N^{-1/3} \times 100 \qquad (1.2)$$

where P_s is the percentage of atoms at the surface of a metal particle and N gives the total number of atoms in the particle [3]

The 3 nm silver nanoparticle we used in the previous example would contain ~1000 atoms [1]. We can therefore estimate that it would have ~40% of the total number of

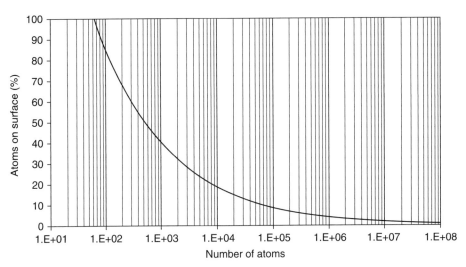

Figure 1.2. A plot of the total number of atoms in a metal particle vs. the percentage of those atoms that are located at the surface of the particle, derived from Equation 1.2.

atoms on the surface. A particle 150 nm in diameter would contain $\sim 10^7$ atoms, with less than 1% of the atoms in surface locations. As heterogeneous catalysis occurs only at the surface of the metal being used, it is obvious to see why the use of nanoparticles is advantageous in order to make optimum use of often expensive catalytic metals.

Aside from these simple surface/volume effects, there are changes in the reactivity of nanoparticles that arise from the quantum confinement effect. This qualitative change in the electronic structure can cause unusual catalytic properties in the nanoparticles that are quite different from the behavior of the bulk material. Photoemission spectroscopy investigations have shown that the electronic structure of metal clusters smaller than ~5 nm is different from that of the bulk metal [4]. The small number of atoms involved in the formation of the electron bands results in a greater localization of the valence electrons and in a smaller width of the valence band. This altered electronic structure and the force towards the center generated by the surface curvature of small metal nanoparticles give rise to a significant contraction of the lattice compared to the bulk material [5]. In turn, the smaller lattice constant is responsible for a shift of the center of the d-band to higher energies, which generally enhances the reactivity of the surface towards adsorbates.

There is also a dramatic increase in the number of edge and corner sites in the metal lattice—and these can react quite differently to the flat metal surface in terms of catalysis properties. The enhancement of the reactivity of low-coordination defect sites can be so large that their presence determines the catalytic activity of a material, in spite of their low concentration [4]. Gold provides an excellent example of a material that behaves markedly differently in the form of nanoparticles. Perhaps the largest breakthrough in nanoparticle catalysis was the work done by Haruta on gold nanoparticle-catalyzed CO oxidation by O_2 at low temperatures [6–8]. Gold is widely recognized as being chemically inert. It is indeed one of the most stable metals in the group 8 elements and is resistant to oxidation. The discovery that gold nanoparticles supported on Co_3O_4, Fe_2O_3, or TiO_2 were highly active catalysts for CO and H_2 oxidation, NO reduction, water-gas shift reaction, CO_2 hydrogenation, and catalytic combustion of methanol was therefore a surprise and considered important by the chemical community [9]. Smooth surfaces of metallic gold do not adsorb CO, and this is necessary for catalysis at room temperatures [10]. CO is instead only adsorbed at corner, step, and edge sites, indicating that smaller metallic gold nanoparticles are preferable [11].

Nanoparticles of a large range of transition metals and metal oxides have been found to exhibit advantageous size-dependent catalytic properties and are being investigated intensively. The shape, coordination, and stabilization of these nanoparticles have been found to affect the catalysis and are therefore also the subject of much current research. As with many other applications of nanoparticles, catalysis often requires a suitable support/substrate for the nanoparticles. This should ideally provide a convenient means to utilize the nanoparticles while protecting them from aggregation and allowing simple recovery. There is therefore much interest in finding effective methods of producing supported nanoparticle catalytic materials using substrates such as inorganic oxides, alumina, silica, and titania, as well as polymers.

1.4 OPTICS AND PHOTONICS

Nanomaterials interact with light differently from the bulk material. Materials in the micro-/nanosize range are of dimensions that are comparable to, or smaller than, the wavelength of light. If a material is of dimensions close to the wavelength of light and is surrounded by a substance of different refractive index, then light of appropriate wavelength will be scattered. The specific wavelength of light that is scattered is dependent on the thickness of the scattering phase. It is this effect that causes oil stretched thinly across the surface of water to produce rainbow colors. This effect has been used in optical materials known as photonic crystals, which are designed with phases of different refractive indices and specific dimensions and architectures intended to produce a desired interaction with light.

In the case of materials in which the separate phase is significantly smaller than the wavelength of light, this effect does not occur. Instead, the two phases behave as a single material with respect to the transmitted light. Therefore, transparent materials with embedded nanoparticles may still remain transparent to light even if the material the nanoparticles are formed from is opaque or reflective in its bulk form. Composites, transparent materials, and inorganic particles in the micrometer range, on the other hand, are often opaque. The light scattering, which is responsible for the opacity, is suppressed either by using materials with nearly matching refractive indices or by decreasing the dimensions of the filler to a range below c. 50 nm. Therefore, nanocomposites with embedded nanoparticles can act as homogeneous materials with modified properties. Instead of scattering light, a merging of the refractive indices of the nanoparticles and host material takes place. Nanoparticles with a high refractive index can be dispersed in a glass or polymer to increase the effective refractive index of the medium.

This approach is helpful in producing optical waveguides where a higher refractive index leads to better beam confinement [12].

Nanoparticles of conducting and semiconducting materials interact directly with light through different mechanisms. It is because of these properties that these nanoparticles are often added to an optical substrate to perform a desired function. Conducting metallic nanoparticles interact with light via an effect known as plasmon resonance, arising from the delocalized cloud of electrons associated with the particle. Semiconductor nanoparticles, often known as "quantum dots," interact with light according to exciton mechanisms modified by the quantum confinement effect. These will be dealt with separately. Quantum dots will only be described briefly as they are not directly relevant to the research reported later in this chapter but are included for comparison.

1.4.1 Quantum Dots

The most significant electronic effect in semiconductor nanoparticles is the widening of the gap between the highest occupied electronic states (at the top of the original valence band) and the lowest unoccupied states (the bottom of the original conduction band). This occurs through quantum confinement because of the small dimensions of the particle [13], which directly affects the optical properties of semiconductor nanoparticles as compared to the bulk material. The minimum energy needed to create an electron–hole pair in a semiconductor nanoparticle (an "exciton") is defined by its band gap (E_g). Light with energy lower than E_g cannot be absorbed by the nanoparticle; the onset of absorption is also size dependent. As the size decreases, the absorption spectra for smaller nanoparticles are shifted to shorter wavelengths; that is, the gap increases in size, with respect to larger nanoparticles or the bulk material.

Excitons in semiconductors have a finite lifetime because of recombination of the photoexcited electron–hole pair. In semiconductor nanoparticles, the energy released by this recombination is too large to be dissipated by vibrational modes. Instead, it is emitted in the form of a photon of appropriate energy. The energy for this fluorescence is centered at a value smaller than the energy required to generate the exciton. If narrow size distributions of quantum dots can be obtained, then a wide range of colors can be produced, even for the same semiconductor material, with each color associated with a different size.

1.4.2 Plasmons

Metallic nanoparticles can have absorption spectra with an absorption peak that looks similar to that of semiconductor nanoparticles. However, this absorption does not derive from transitions between quantized energy states; instead, in metal nanoparticles, collective modes of motion of the electron cloud can be excited [14]. Under the influence of an electrical field, there is a plasmon excitation of the electrons at the particle surface. This resonance takes place at a certain frequency of incident light and results in an optical absorption. These are referred to as surface plasmons, or also as plasma resonance absorption, or as localized surface plasmons.

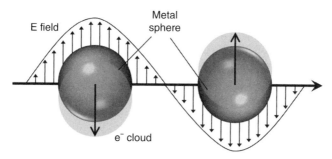

Figure 1.3. Schematic of plasmon oscillation for a sphere, showing the displacement of the conduction electron charge cloud relative to the nuclei, as described by Kelly et al. [17].

As particle size decreases, the conduction electrons begin to interact with the boundary of the particle [15]. A metal–dielectric boundary on the nanoscale produces considerable changes in the optical properties, making them size and shape dependent [16]. When a metallic nanoparticle is irradiated by light, the electric field of the incident light induces a coherent oscillation of the conduction electrons [17] (see Fig. 1.3). For nanoparticles significantly smaller than the wavelength of the light, absorption is within a narrow wavelength range, a plasmon band.

The width, position, and intensity of the plasmon interaction displayed by nanoparticles depend on [1, 16–19]:

- The dielectric functions of the metal and the host material
- Particle size and shape
- The interface between the particle and the host
- The distribution of particles within the host

Because of the influence of these factors, a desirable amount of control over the final properties of the material is possible, provided that an equal amount of control is possible in modifying the factors themselves. Different metals produce different light interactions and therefore colors. The extinction of light by metal nanoparticles occurs by both scattering and absorption mechanisms, but absorption is by far the dominant factor for nanoparticles of small size (<20 nm). Nanosized metal particles have long been used for coloring glass from Roman times and through the Middle Ages in church windows. Although this particular application of nanoparticles may have been unintentional, small nanoparticles were often used to create glass with ruby-red Au and lemon-yellow Ag nanoparticles.

Today most of the use and study focuses on gold and silver nanoparticles because they exhibit the most pronounced plasmon effects and both have absorptions in the visible spectrum. Increasing the size of the particles, or increasing the dielectric constant of the medium, causes a redshift in the plasmon absorption [16]. However, it must be noted that the effect of nanoparticle size, though still significant, is considerably less dramatic than that observed in quantum dots. The absorption peak position in quantum dots is shifted markedly for a change in diameter of only fractions of a nanometer. For metal nanoparticles the shift in peak position is minimal for small particles (<25 nm for gold), although a broadening does take place. For larger metal particles (>25 nm for gold), the redshift of the plasmon resonance position is more significant [16]. If the particles are distorted in shape, then the plasmon band splits into different modes corresponding to the different aspects of the electron oscillations. For example, in a nanorod-shaped metallic nanoparticle, the plasmon band splits into two bands corresponding to oscillation of the free electrons along (longitudinal) and perpendicular (transverse) to the long axis of the rod. The transverse mode resonance is close to that observed for spherical particles, but the longitudinal mode is redshifted, dependent on the aspect ratio of the nanorod [16, 19].

Metal nanoparticles of this nature, intended for use in optical or photonic applications, are often embedded into a host material such as an appropriate polymer or glass. The incorporation of metal nanoparticles into optical substrates allows the construction of devices to utilize their advantageous properties. The host matrix not only forms the structure of the device but also protects the nanoparticles and prevents agglomeration. There are widespread uses for these emerging materials especially in optical and photonic applications including eye and sensor protection, optical communications, optical information processing, Raman enhancement materials, optical switching, plasmon waveguides, light-stable color filters, polarizers, and modified refractive index materials.

Silver nanoparticles are often singled out as an ideal choice of optically active nanoparticles for use in such research. Although silver and gold nanoparticles share many similar properties and applications, suitable silver feedstocks remain considerably more economically viable. In addition, silver represents the material with the most, and more general, applications. The reason for the considerable interest in the use of silver nanoparticles can best be summarized by reviewing the following statement from Reference [20]:

Of the three metals (Ag, Au, Cu) that display plasmon resonances in the visible spectrum, Ag exhibits the highest efficiency of plasmon excitation. Moreover, optical excitation of plasmon resonances in nanosized Ag particles is the most efficient mechanism by which light interacts with matter. A single Ag nanoparticle interacts with light more efficiently than a particle of the same dimension composed of any known organic or inorganic chromophore. The light-interaction cross-section for Ag can be about ten times that of the geometric cross-section, which indicates that the particles capture much more light than is physically incident on them. Silver is also the only material whose plasmon resonance can be tuned to any wavelength in the visible spectrum. (Reference 20, 1222. Copyright Wiley-VCH Verlag GmbH & Co. KGaA. Reproduced with permission.)

From this statement the obvious advantages of silver, and the reason for its use as a model nanoparticle for the synthesis of nanocomposite materials, are clear. Noble metal nanoparticles have many applications; three major opportunities of these are outlined briefly in the succeeding text to demonstrate how these strong plasmon interactions can be utilized.

1.4.3 Nonlinear Optical Limitation

Many applications of optical composites of silver or gold nanoparticles derive from their potential as optical limiters. This arises from their nonlinear susceptibility near the surface plasmon frequency with picosecond response time [21]. Optical limitation is the decrease in absolute transmittance of a material as incident light fluence (energy per unit area) increases. Optical limiting materials have received considerable attention as a result of a growing need for optical sensor protection. The advance of optical detection systems for applications such as signal acquisition and night vision demands the development of protective devices, such as passive optical limiters, so that highly sensitive detectors can survive undesired high-intensity signals. Also, the ever-increasing use of lasers calls for more versatile and improved forms of eye protection. An ideal optical limiter should display the capability of being transparent at low fluences of incident light and opaque at high fluences [22]. It should also be capable of switching back and forth between these two states in quick response to external optical signals. However, most materials do not possess such an unusual property; instead, they often become more transparent under a strong light signal owing to the depletion of the electronic ground state. Silver nanoparticles, on the other hand, have been found to function as optical limiters by reverse saturable absorption. The basis of this effect is the promotion of electrons from a weakly absorbing ground state to a more strongly absorbing excited state. It has been proposed that this behavior is the result of photoinduced intraparticle charge separation leading to strong free-carrier absorption [23]. While most metal nanoparticles exhibit a weak optical limitation, it has been found to be particularly strong for gold and silver [24].

Sun et al. [23] prepared a stable suspension of silver nanoparticles and compared the optical limitation effects to similar suspensions of C_{60} and chloroaluminum phthalocyanine, which are normally considered as benchmark materials for high-performance optical limitation. The silver nanoparticles were found to be significantly more effective. This study, as well as more recent research [25, 26], shows the great potential of silver nanoparticles for use in optically limiting devices, especially if they can be effectively incorporated into transparent polymer supports [27].

1.4.4 Surface-Enhanced Raman Spectroscopy

The processes of infrared absorption and Raman scattering are routinely used in spectroscopy to detect vibrations in molecules. This can provide information on chemical structure and so helps to identify molecular structures and to determine compositions of substances and mixtures. Recent advances in instrument technology and the ability of Raman spectroscopy to examine aqueous solutions and samples inside glass containers without any preparation have led to a rapid growth in the application of the technique [28].

Raman spectroscopy relies on inelastic scattering, or Raman scattering of monochromatic light, usually from a laser in the visible, near-infrared, or near-ultraviolet range. The laser light interacts with phonons (quantized modes of vibration) resulting in the energy of the laser photons being shifted up or down. The shift in energy gives information about the vibrational modes in the system. Infrared spectroscopy yields similar but complementary information. Surface-enhanced Raman spectroscopy (SERS) uses silver or gold nanoparticles or a substrate containing silver or gold. The surface plasmon resonances of silver and gold are easily excited by the laser, and the resulting enhanced electric fields around the nanoparticles cause other nearby molecules to become Raman active. The result is amplification of the Raman signal, making it possible to detect very small concentrations of analytes. The particle enhances not only the incident laser field but also the Raman scattered field. It acts as an antenna that amplifies the scattered light intensity. The reason that the SERS effect is so pronounced is because the field enhancement occurs twice. Initially, the field enhancement magnifies the intensity of incident light that will excite the Raman modes of the molecule being studied, therefore increasing the signal of the Raman scattering. The Raman signal is then further magnified by the surface according to same mechanism, resulting in a greater increase in the total output signal of the experiment. With enhancement factors of the order of 10^{14} possible, this makes SERS an extremely sensitive spectroscopic technique to the extent that applications can extend down to single molecule detection [29]. This has created a drive to develop synthesis methods for effective metal nanoparticle-based SERS materials that can be easily applied to practical analysis.

1.4.5 Metal-Enhanced Fluorescence

Metal-enhanced fluorescence (MEF), a phenomenon where the quantum yield and photostability of weakly fluorescing species are dramatically increased, is becoming a powerful tool for the fluorescence-based applications of drug discovery, high-throughput screening, immunoassays, and protein–protein detection [30]. Fluorescence has become the dominant detection/sensing technology in medical diagnostics and biotechnology. Although fluorescence is a highly sensitive technique, there is still a drive for reduced detection limits. The detection of a fluorophore is usually limited by its quantum yield, autofluorescence of the samples, and/or the photostability of the fluorophores. However, there has been a recent upsurge in the use of metallic nanostructures to favorably modify the spectral properties of fluorophores and to alleviate some of these fluorophore photophysical constraints. The use of fluorophore–metal interactions has been termed radiative decay engineering, MEF, or surface-enhanced fluorescence [31].

Silver nanoparticles deposited on substrates have been widely used for a variety of applications in MEF typically producing >5-fold enhancements in emission intensity [31]. Metal-enhanced fluorescence is a through-space phenomenon. When fluorophores are placed within 4–10 nm of the surface nanostructures, the emission of the fluorophores is enhanced [31]. The increased emission intensity is a result of the metal nanoparticles increasing the local incident field on the fluorophore.

1.5 GENERAL SYNTHETIC STRATEGIES

The methods of producing nanoparticles are commonly separated into two main categories, "top-down" and "bottom-up" approaches.

1.5.1 Top Down

The top-down method involves the systematic breakdown of a bulk material into smaller pieces usually using some form of grinding mechanism. This is advantageous in that it is simple to perform and avoids the use of volatile and toxic compounds often found in the bottom-up techniques. However, the quality of the nanoparticles produced by grinding is widely accepted to be poor in comparison with the material produced by modern bottom-up methods. The main drawbacks include contamination problems from grinding equipment, low particle surface areas, irregular shape and size distributions, and high energy requirements needed to produce relatively small particles. Aside from these disadvantages, it must be noted that the nanomaterial produced from grinding still finds use, due to the simplicity of its production, in applications including catalytic, magnetic, and structural purposes.

1.5.2 Bottom Up

The bottom-up approach uses atomic or molecular feedstocks as the source of the material to be chemically converted into larger nanoparticles. This has the advantage of being potentially much more controllable than the top-down approach. By controlling the chemical reactions and the environment of the growing nanoparticle, then the size, shape, and composition of the nanoparticles may all be affected. For this reason, nanoparticles produced by bottom-up, chemically based and designed, reactions are normally seen as being of higher quality and having greater potential for use in advanced applications. This has led to the development of a host of common bottom-up strategies for the synthesis of nanoparticles.

Many of these general techniques can be adapted to be performed in gas, liquid, solid, or even supercritical states, hence the applicability of bottom-up strategies to a wide range of end products. Most bottom-up strategies require appropriate organometallic complexes or metal salts to be used as chemical precursors that are decomposed or reduced in a controlled manner resulting in particle nucleation and growth. One of the key differences that can be used to subdivide these strategies into different categories is the method by which the precursor is decomposed or reduced. It is beyond the scope of this book to describe all current and historical bottom-up synthetic methods of nanoparticles, as there are a great number of variations. However, as the majority of methods used involve solution synthesis based on a few well-known general techniques, the most common and applicable of these are briefly described.

1.5.3 Solution Synthesis

Scientific synthesis of nanoparticle solutions can be said to have started with experiments begun by Michael Faraday on gold in the mid-nineteenth century. Deep-red solutions of gold nanoparticles were produced by the reduction of chloroaurate $[AuCl_4]^-$ using phosphorus as a reducing agent. This is an example of chemical reduction of the metal precursor, the most common method for the generation of the nanoparticle-forming material. Other methods include thermal decomposition and photoreduction of metal ions [32–34]. For routes in which the precursor organometallic complex or metal salt is chemically reduced, this can be accomplished by using a reducing solvent such as alcohols as pioneered by Hirai and Toshima [35] or by using an added reducing agent dissolved or otherwise introduced into a nonreducing solvent. Gold nanoparticles are now routinely produced in solution, to a monodisperse particle size, by the chemical reduction of $AuCl_4^-$ by sodium borohydride in the presence of alkanethiol stabilizing agents [36]. The majority of the more straightforward approaches used for the synthesis of silver nanoparticles are based on the reduction of silver nitrate by sodium borohydride [37] or sodium citrate [38]. Other, more diverse, methods of decomposing the precursor species include microwave plasma synthesis and electrolysis of metal salts. In many nanoparticle synthesis routes, organometallic precursors play an important role by allowing more control through tailored design; that is, by selecting appropriate ligands, the solubility and decomposition rate of the precursor can be modified.

1.6 STABILIZATION

No matter how they may have been created, nanoparticles generally require some form of stabilization to prevent them from coalescing, agglomerating, or aggregating—which can detrimentally affect their properties and application. Metal oxide nanoparticles are more stable, but unprotected metal nanoparticles are subject to strong attractive forces especially at short interparticle distances. The nanoparticles are attracted together by van der Waals forces and, because of their metallic lattice structure, can easily coalesce. One method of stabilizing nanoparticles is simply to deposit, infuse, or embed the nanoparticles into a solid substrate/host material. This matrix must, of course, be suitable for the application that the nanoparticles are intended for. However, if the nanoparticles are required for synthesis or use in solution, then other techniques must be used. The two possible routes for the stabilization of nanoparticles in solution are electrostatic stabilization and steric stabilization.

1.6.1 Electrostatic Stabilization

This method is often used in the more traditional routes to nanoparticle synthesis, especially for well-controlled particle size distribution. Each metallic nanoparticle is surrounded by an electrical double layer, which causes repulsion between neighboring nanoparticles. This electrical double layer forms because of the attraction to the surface of the nanoparticle of negative ions present in the solution. These negative ions are attracted to positive metal ions at the surface of the metallic lattice of the nanoparticle. This helps to stabilize and control the growth of the forming nanoparticle. The negative ions species can be by-products of the metal feedstock (the metal salt), the reducing agent, or just species present in the solution. This coulombic repulsion between the particles caused by the electrical double layer formed by ions adsorbed at the particle surface (e.g., sodium citrate) and the corresponding counterions is exemplified by gold

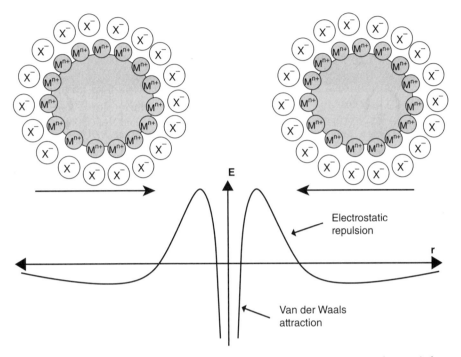

<u>Figure 1.4.</u> Electrostatic stabilization of metal nanoparticles. Attractive van der Waals forces are outweighed by repulsive electrostatic forces between adsorbed ions and associated counterions at moderate interparticle separation.

sols prepared by the reduction of $[AuCl_4]^-$ with sodium citrate [39]. See Figure 1.4 for a schematic representation of this process.

1.6.2 Steric Stabilization

Steric stabilization is achieved by the coordination of sterically demanding molecules such as polymers, surfactants, or ligands that act as protective shields on the metallic surface. In this way, nanometallic cores are separated from each other and agglomeration is prevented [40] (see Fig. 1.5). This can occur because of the electron-deficient nature of the metal surface. Metal ions in the bulk of a lattice are commonly surrounded on all sides by the delocalized electrons in the structure. This leaves the metallic ions on the lattice surface, especially the curved surface of a nanoparticle, comparatively electron deficient. Therefore, it is necessary for steric protecting agents to have suitable electron-donating groups. These groups coordinate to the surface of the metal nanoparticle.

The main classes of protective groups are polymers and block copolymers, usually with P, N, O, and S donors (e.g., phosphanes, amines, esters/ethers, thioethers), or solvents such as tetrahydrofuran and methanol that have electron-rich groups. The steric protecting agent, in order to function effectively, must not only be attracted to the nanoparticle surface but also be adequately solvated by the dispersing fluid.

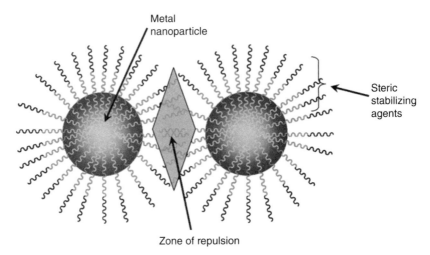

Figure 1.5. Steric stabilization of nanoparticles. Stabilizing agents may be uniform or may be composed of different segments to provided separate functions, for example, a nanoparticle-philic segment and a solvent-philic segment.

1.7 POLYMERS

From the earliest recorded times, man has exploited naturally occurring polymers as materials for providing clothing, shelter, and tools. Nowadays synthetic polymers have considerable commercial importance and have become an integral part of our daily lives. From the developments of the first commercial synthetic polymers in the early twentieth century to today, the breadth of polymer science has been ever increasing, and regular advances have continued to stimulate both scientific and industrial progress. The use of polymeric materials is increasing rapidly year by year as polymers are tailored to replace conventional materials such as metals, wood, and natural fibers. The degree to which polymer properties can now be adapted and controlled by modification of chemical structure, the blending of polymers, or the production of unique architectures has led to polymeric materials being designed to fill new and specific roles that no other known material could provide. It is because of this need for new functional materials that the development of polymeric materials has become combined with the emergent field of nanotechnology via the incorporation of nanomaterials into polymers to produce novel hybrid materials.

1.7.1 Definition

Polymers are large molecules made up of repeating units. They are synthesized from simple monomer molecules. These monomers become linked together forming chains increasing in size from dimers and trimers to oligomers and eventually polymers with a molecular weight of at least several thousand atomic mass units or more. By changing from a monomeric to a polymeric structure, the nature of the material is greatly altered.

While the monomeric substance may have been naturally liquid or even gaseous, it will become a solid or at least a viscous liquid as a polymer. This is because the polymeric chains become entangled with each other, their molecular forces of attraction to each other are greatly compounded, and the timescale of movement of the repeating units is severely retarded. It is useful to describe some general features of polymers here, such as crystallinity and glass transition. The production of metal nanoparticle–polymer composites will also be given a general overview.

1.7.2 Crystallinity in Polymers

The long chains that make up the structure of the polymers exist in one of two possible arrangements:

- Amorphous—chains form a randomly ordered, chaotic "mess" (like a plate of spaghetti).
- Crystalline—chains are arranged in an ordered manner (like dry spaghetti in a packet).

It is normal for these arrangements to coexist in the bulk of a polymer, with the proportions of each depending on the polymer in question. Most polymers are therefore described as semicrystalline (Fig. 1.6).

Crystallinity makes a polymer denser, more resistant to solvents, and more opaque (the crystallites scatter light). The mechanical properties of a polymer are also affected by its crystallinity as well as the position of its glass transition temperature.

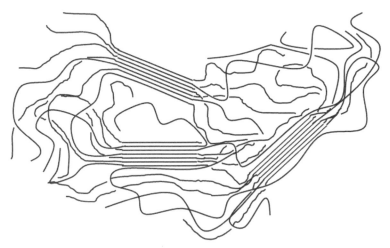

Figure 1.6. Schematic of semicrystalline polymer showing ordered crystallites separated by amorphous areas.

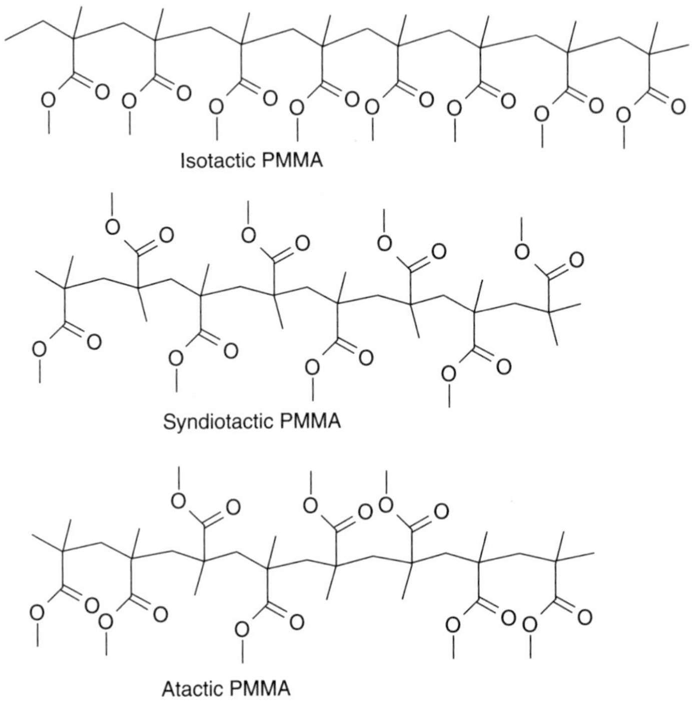

Figure 1.7. Possible tacticities of PMMA. Which tacticity is formed is governed by the method of polymer preparation. The free radical polymerization mechanism most commonly used allows no control over the position the side group takes and so produces atactic PMMA.

Poly(methyl methacrylate) (PMMA) provides a good example for discussing how the structure of the individual polymer chains of a polymer can determine the crystallinity of the material. When PMMA has an atactic structure (Fig. 1.7), the lack of order of the side groups prevents the chains from being able to pack together effectively, leading to a disordered, amorphous material. Isotactic PMMA, alternatively, is quite crystalline. Its ordered structure allows the polymer chains to effectively pack together into crystallites.

1.7.3 The Glass Transition and Melting Point

The glass transition occurs at a specific temperature, T_g, that is different for each polymer. Below the glass transition temperature, the polymer is hard and brittle (like glass); above, it is softer and more flexible (like rubber). This glass transition is very different to melting. Melting (T_m) occurs in the crystalline part of polymers, when rigid chains in the crystals loosen, becoming a disordered liquid. The glass transition is something inherent to only the amorphous region. It is therefore normal for polymers to have both a T_g and a T_m if they possess amorphous and crystalline parts.

The position of the T_g is very important in polymer engineering and use. It governs how easy they are to process and what use they can be put to. Whether a polymer's T_g is above or below room temperature, or close to it, will affect the possible end applications of the material. This transition is related to the amount of mobility that the polymer chains possess. Below the T_g the chains have relatively little scope for free movement, though the segments are still moving and rotating a good deal. Above the T_g the chains have gained a greater mobility, and their thermal energy causes increased movement, creating more free volume for the polymer chains to slide past each other more easily. Because of this, if a force is applied to a polymer that is above its T_g, the chains are able to move to relieve the stress, and the polymer bends. If a force is applied to a polymer that is below the T_g, the chains will be unable to move in this way, and the polymer will either resist the force or break.

1.8 METAL–POLYMER NANOCOMPOSITES

The embedding of nanoscopic metals into dielectric matrices represents a solution to manipulation and stabilization problems. For functional applications of nanoparticles, polymers are particularly interesting as an embedding phase since they may have a variety of characteristics: they can be an electrical and thermal insulator or conductor, may have a hydrophobic or hydrophilic nature, can be mechanically hard (plastic) or soft (rubbery), and so on. Finally, polymer embedding is the easiest and most convenient way for nanostructured metal stabilization, handling, and application. This has fuelled investigation into the preparation of metal–polymer nanocomposites. These composites most commonly take the form of thin polymer films or powders, as this is normally the simplest structure to prepare and also good for exploiting the desired properties. Preparation techniques can be classified as *in situ* and *ex situ* methods. In the *in situ* methods, the monomer is polymerized, with metal ions introduced before or after polymerization. Then the metal ions in the polymer matrix are reduced chemically, thermally, or by UV irradiation, to form nanoparticles. In the *ex situ* process, the metal nanoparticles are synthesized first, and their surface is organically passivated. The derivatized nanoparticles are then dispersed into a polymer solution or liquid monomer that is then polymerized. A more detailed description of some of these key techniques follows.

1.8.1 *Ex Situ*

The metal nanoparticles are prepared first, traditionally by the controlled precipitation and concurrent stabilization of the incipient colloids. This can be done by the reduction of a metal salt dissolved in an appropriate solvent, often containing a polymer stabilizer. Alternatively, it can be prepared by controlled micelle, reverse micelle, or microemulsion reactions. The particles produced by these methods are often surface modified to prevent aggregation, either covalently by metal–thiol bonds [41] or by coating with a suitable polymer shell [42]. These particles then need to be introduced into polymers. This is accomplished by mixing with a solution of the polymer, or monomer, which can then be spin cast according to standard polymer processing techniques. However, this method is limited by problems of dispersion. It is necessary to surface modify the particles, therefore altering their properties, in order to disperse them. Even with this step, it is difficult to produce well-dispersed composites, and a certain degree of aggregation remains. Also, this route is limited to compatible polymer–particle–solvent systems.

1.8.2 *In Situ*

The *in situ* methods that have been used for the manufacture of metal nanoparticle–polymer composites consist of much more varied techniques. *In situ* methods, though often less simple and straightforward as *ex situ*, are commonly considered to produce better quality and more controlled nanocomposite materials. An outline of the most significant methods follows.

1.9 THERMAL DECOMPOSITION OF METAL PRECURSORS ADDED TO POLYMERS

This represents the existing method most suited to large-scale production. A number of organic precursors were studied for this application without showing completely satisfying behavior until it was discovered that homoleptic mercapeptides (i.e., $M_x(SR)_y$) are effective in the production of metal–polymer nanocomposites [43]. Transition metal mercapeptides are covalent organic salts with a high compatibility with most hydrophobic polymers. The mercapeptide is dispersed in a polymer, and the polymer is then heated to 110–180 °C to decompose the mercapeptide. This produces zerovalent metal or metal–sulfide nanoparticles, depending on the metal and conditions. The literature also reports similar interesting composites prepared by the inclusion of metal nitrates (e.g., $AgNO_3$) [44–47] and metal salts (e.g., $HAuCl_4$) [48]. In these methods organic by-products are left trapped in the polymer, and the reduction of the metal causes damage to the substrate by electron extraction. With any nanocomposites produced by this method, the precursors must be included in the polymer *before* it is cast into a solid form. Any subsequent processing may then damage or affect the composite.

1.10 ION IMPLANTATION

An ion beam (e.g., Ag^+) of the range 30–150 keV in energy is directed at the surface of a polymer sample. As the ions enter the matrix, nuclear collisions occur, displacing atoms in the polymer matrix and breaking some of its chemical bonds. Along with this, target atoms effectively lose electrons, and the implanted M^+ ions deionize with the formation of neutral metal atoms (M^0). It is possible in principle for metal atoms to combine with organic radicals and polymer ions. However, because of the great difference in Gibbs free energy between metal atoms and atoms of the polymer, metal–metal bonding is energetically more favorable. Though offering good control of particle size and dispersion, this method is limited to impregnating a few micrometers into the surface. Also, it can lead to the formation of carbonized shells around the metal nanoparticles. Nevertheless, it has been used for the effective synthesis of metal–polymer nanocomposite materials including PMMA–Ag nanocomposites [49, 50].

1.11 CHEMICAL VAPOR DEPOSITION (CVD) AND PHYSICAL VAPOR DEPOSITION (PVD)

These methods, though consisting of many subdivisions, involve the formation of a solid from a controlled deposition of gaseous species. Both processes are commonly used, especially in the electronics industry, for coating materials with thin films but have also been used to produce nanoparticles. Due to the nature of these processes, the nanoparticles are commonly deposited onto a suitable substrate, such as an inorganic support, or with a co-deposited polymer.

Chemical vapor deposition (CVD) is a chemical process used to produce high-purity, high-performance solid materials. In a typical CVD process, the substrate is exposed to one or more volatile precursors, which react and/or decompose on the substrate surface to produce the desired deposit. Frequently, volatile by-products are also produced that are removed by gas flow through the reaction chamber. The precursors used are volatile organometallic complexes, similar to those sometimes used in the synthesis of nanoparticles through solution methods. In CVD, precursor gases (often diluted in carrier gases) are delivered into the reaction chamber at approximately ambient temperatures. As they pass over or come into contact with a heated substrate, they react or decompose forming a solid phase that is deposited onto the substrate. Okumura et al. used this technique to form catalytic nanoparticles of gold on inorganic substrates [51].

Physical vapor deposition (PVD) is similar but allows the nanocomposite films to be generated more simply from the pure material/materials intended to form the nanoparticles/composite. This is achieved by thermal evaporation (by an electron beam or resistive heating) of the starting materials to form a vapor phase. A carrier gas is then introduced to transport the evaporated metal away from the heater, towards the cool end of the reactor. As the vapor cools, nucleation occurs and liquid metal droplets form. These droplets collide, coalesce, and solidify in a manner controlled by the temperature gradient, residence time, and gas flow rates employed during the procedure. This is preferable to CVD as it produces no by-products. Oxygen is frequently used as a carrier gas in order to produce metal oxide particles.

When these methods are used for the simultaneous synthesis of metal nanoparticles and a polymer substrate, this forms the nanoparticles inside the polymer substrate with a good deal of control on the structure of the composite produced being possible. By controlling the flows of the two feedstocks, the nanocomposite can be made continuous, gradated or layered, etc.

1.12 scCO₂ IMPREGNATION INTO POLYMERS

Supercritical fluid (SCF) methods for the generation of metal nanoparticles in solid polymers, by infusing precursor complexes subsequent to their reduction, are outlined in the following section. The advantages of the scCO₂ method are many. It is a green process, avoiding the use of toxic and harmful solvents as well as the problems of their removal. It allows the modification of polymers that are difficult to process by normal methods. It has promising potential for effective and economical scale-up. Importantly, it allows the production of pure, non-surface-coated nanoparticles, without leaving any trapped by-products in the polymer. It would also allow prefabricated polymeric materials to be impregnated with nanoparticles after they had been processed into their final form, which avoids detrimental aggregation of the nanoparticles during processing.

1.13 SUPERCRITICAL FLUIDS

1.13.1 The Discovery and Development of SCFs

Supercritical fluids are useful because they possess both liquid-like and gas-like properties. The critical point of a substance was first discovered and defined by *Baron Charles Cagniard de la Tour* in 1822. His experiments proved the existence of SCFs by heating various solvents sealed within metal cannons. By observing the changes in solvent acoustics, *de la Tour* was able to identify the existence of a *new* single phase, rather than the two separate liquid and gas phases [52].

This initial discovery sparked subsequent investigations, and in 1869 Dr. Thomas Andrews reported the critical parameters of carbon dioxide and illustrated the phases by constructing the first supercritical phase diagram [53]. In 1879, at a *Royal Society* meeting, *Hannay* and *Hogarth* reported the ability of certain compressed gases to dissolve salts such as cobalt chloride, potassium iodide, and potassium bromide, which was previously thought to be impossible [54]. In summary:

> We have the phenomenon of a solid dissolving in a gas, and when the solid is precipitated by reducing the pressure, it is brought down as a "snow" in the gas or on the glass as a "frost," but it is always easily redissolved by the gas on increasing the pressure (326).

These initial investigations lead on to further work and the eventual definition of an SCF, by Darr and Poliakoff, as "*any substance, the temperature and pressure of which are higher than their critical values, and which has a density close to or higher than its critical density*" [55]. This is best explained with reference a phase diagram (Fig. 1.8). Similar to gases, SCFs possess high diffusivities, important for reaction kinetics, while also having liquid-like densities that allow them to act as effective solvents for many compounds. Small changes in pressure and temperature can be used to "tune" the density of an SCF and therefore its solvating ability, hence giving increased control of reactions.

Supercritical fluids are unique solvents with a wide range of interesting properties. They can be clean and versatile, capable of processing a wide variety of materials. An increasing drive for greener processes in chemistry and industry in the 1970s led to a move towards the wider use of SCFs as alternative reaction media, for example, for extraction and separation processes. This marked the need for a move away from volatile organic compounds and ozone-depleting substances such as chlorofluorocarbons, which are often used as processing solvents. Many of these traditional solvents are not only environmentally harmful but often hazardous and toxic to humans, with high costs of waste removal.

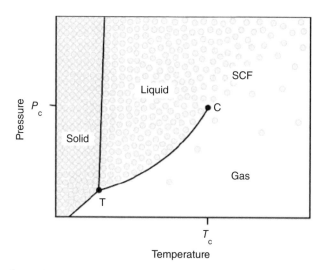

Figure 1.8. Schematic pressure–temperature phase diagram for a pure component showing the SCF region. The triple point (T) and critical point (C) are marked. The circles show the density variation of the substance in the different regions of the phase diagram. Note that the density varies continuously between the liquid state and the gas state, provided that the liquid–gas equilibrium line is not crossed [56]. Reproduced by permission of The Royal Society of Chemistry.

1.13.2 Supercritical CO_2

CO_2 in particular has received much attention as an SCF and is probably the most widely used. The choice of which SCF is used for chemical and industrial processes must be determined by a compromise of practical factors. Supercritical processes have associated costs arising from the need for high-pressure equipment, and therefore solvents with more easily reached critical points will lower this cost, but this must be offset by the financial cost of the solvent, ease of handling, solvent reactivity, and safety factors.

CO_2 has relatively easily accessible critical points of 31.06 °C and 1070 psi (7.38 MPa) and tunable density, as well as a high diffusivity, making it a promising reaction medium for organic compounds. There are of course other substances that share these properties. However, many of them have considerable disadvantages not present for $scCO_2$. Most of the organic solvents suitable for supercritical applications are flammable, often dangerously so. They also tend to have potentially harmful effects to health and the environment. CO_2 is noncombustible and nontoxic as well as being relatively environmentally benign. It is also easily available as it is naturally occurring as well as being the by-product of many industrial processes, such as the burning of fossil fuels for power. As a consequence of this, both high-quality and low-quality stock can be easily obtained, meaning that CO_2 is relatively inexpensive and financially desirable as a reaction medium. Water also shares some of these advantages of nontoxicity and availability. However, high critical parameters and corrosion problems with metal equipment impede its wider application.

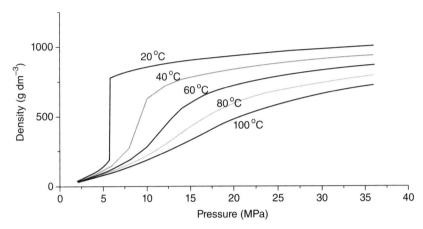

Figure 1.9. The density of CO_2 as a function of pressure for a range of temperatures.

As CO_2 is a gas at ambient temperatures and pressures, it can be easily removed, leaving no solvent residues in the processed material. The ease of its removal contrasts with the high energy costs of drying for the removal of conventional solvents. In small-scale reactions, the CO_2 contribution to greenhouse effects will be negligible, but for larger-scale processes, CO_2 can be recycled back into the reaction system.

The density of $scCO_2$ in a range of temperatures and pressures is shown in Figure 1.9. It is this easily tunable density, and consequently solvent strength, that allows facile control of reactions and processes.

Many of the processes for which $scCO_2$ is used are aided by zero surface tension, allowing hydrophobic behavior normally associated with liquid to substrate interactions, and other phenomena such as drying-induced substrate damage, to be avoided. This has allowed $scCO_2$ to be utilized as an ideal solvent for the cleaning and processing of delicate electronic components and micro-devices. As well as use in microelectronics, $scCO_2$ has found widespread application in other industrial processes, such as the extraction of metals or organic material. CO_2 has also been used for foodstuffs, notably extracting caffeine from coffee, and also for dry cleaning and degreasing. CO_2 is even being employed in increasingly large scales for organic synthetic reactions and transformations [57–59]. However, the two aspects of $scCO_2$ use most relevant to the scope of this book are the processing and synthesis of polymers, and $scCO_2$ generation and impregnation of nanoparticles. These topics will be discussed in greater detail later.

1.14 POLYMER PROCESSING WITH $scCO_2$

The solvating ability of $scCO_2$ when interacting with polymers has been compared to volatile organic solvents, such as hexene. The density of $scCO_2$, as governed by its temperature and pressure, affects its solvent nature, which is commonly measured by dielectric constant. While a good solvent for many polar and nonpolar molecules of low molecular weight (e.g., monomers, initiators, oligomers), $scCO_2$ is a poor solvent for most high-molecular-weight polymers [56]. The solubility of polymers in scCO$_2$ is governed by the CO$_2$–polymer segment interactions, compared to the strength of the segment–segment and CO$_2$–CO$_2$ interactions, and the flexibility of the polymer chain [60]. As a quadrupolar molecule, CO$_2$ does not strongly interact with nonpolar groups and hence, combined with having a lower density than most liquid solvents, is a poor solvent for many long-chain polymers. Solubility in scCO$_2$ generally increases with the number of polar groups in the polymer [61]. However, for very polar polymers, segment–segment interactions will be too strong, therefore reducing solvation.

The factors that control polymer segment–CO$_2$ interactions are not yet fully understood. Fluorine-functionalized polymers have been widely investigated, as they seem to have a greater affinity for CO$_2$. However, this appears to be linked to carbonyl/ether groups present in the polymers, which are generally accepted to have a favorable interaction with CO$_2$. It is noted that poly(tetrafluoroethylene) (PTFE), completely fluorinated as it is, though with no carbonyls/ethers, is no more miscible with scCO$_2$ than a polyolefin. Three important features that help in making a polymer scCO$_2$ soluble are:

1. Low cohesive energy
2. High free volume
3. Groups that provide a specific interaction with scCO$_2$, that is, F, C$=$O, ether etc.

For polymers that are soluble, or semi-soluble, scCO$_2$ can be used for the extraction of low-molecular-weight material, as well as polymer fractionation. Polymerization reactions have been successfully carried out in scCO$_2$, often with surfactants or cosolvents to aid solubility or dispersion of higher-molecular-weight species [56, 62, 63]. These reactions are capable of producing easy to process particulate powders of solid polymer, and there are numerous publications in the literature where scCO$_2$ has been used as a solvent to facilitate clean polymer synthesis [62, 64–73]. Much of this research has centered on the ability to develop CO$_2$-soluble polymer segments, to stabilize the growing polymer chain, enhance its dispersion in scCO$_2$, and ensure controllable, reproducible polymer preparations.

Insoluble polymers have received at least as much attention for use with scCO$_2$. Even though they are not soluble in scCO$_2$, it is still able to permeate into them, leading to some very useful effects. Even a small amount of scCO$_2$ absorbed into a polymer phase results in substantial and sometimes dramatic changes in the physical properties that dictate processing. These include viscosity, permeability, interfacial tension, and glass transition temperature. The permeation of scCO$_2$ into a polymer causes it to swell in volume. The CO$_2$ molecules, aided by their zero surface tension, permeate easily into a polymer matrix, allowing the chains a greater mobility. The CO$_2$ molecules act as "molecular lubricants" between the chains, giving them more free volume of movement and therefore reducing the T_g. This also gives the polymer chains the freedom to align themselves into a more favorable order, hence increasing crystallinity.

The permeation of scCO$_2$ into polymers has been used for extraction, foaming, and impregnation. Extraction occurs by simply removing the soluble extractant material, such as unreacted monomer, while leaving the insoluble substrate. Foaming occurs when rapid decompression forms gaseous CO$_2$ inside the polymer. As for impregnation, by utilizing its properties as a transport medium, molecules can be dissolved in $scCO_2$, impregnated into a polymer, and CO_2 can be cleanly removed afterwards. Substances impregnated into polymers have included dyes, fragrances, drugs for controlled release, antimicrobial and antifungal agents, and metal nanoparticles [74]. The impregnation technique has also led to a novel method of producing polymer blends by introducing a monomer before initiating polymerization.

1.15 NANOPARTICLES BY scCO₂ IMPREGNATION OF HOST MATERIALS

Supercritical fluids have been used to deposit thin metal films onto a wide range of surfaces and to incorporate metallic particles into different inorganic and organic substrates for microelectronic, optical, and catalytic applications [75]. This technique has allowed both highly dispersed and uniformly distributed metal nanoparticles and agglomerated clusters of nanoparticles of wider size distribution to be generated in host materials. These various solid substrates have included silicon wafer, metal foil, inorganic nanotubes, bulk polymers and polymer membranes, and organic and inorganic porous materials. The nanoparticle impregnation process involves dissolving an appropriate metallic precursor in an SCF (most commonly $scCO_2$) and then exposing the substrate to the solution. The SCF facilitates the infusion of the precursor into the host material. After incorporation of the precursor with the substrate, the metallic precursor is reduced to its metal form by a wide variety of methods resulting in films or particles. The reduction methods employed are chemical reduction in the SCF with a reducing agent such as hydrogen and alcohols, thermal reduction in the SCF, and thermal decomposition in an inert atmosphere or chemical conversion with hydrogen or air (Fig. 1.10).

The use of an SCF such as $scCO_2$ for this impregnation process is advantageous in comparison to conventional solvents for many reasons. For many of the substrates used, a conventional solvent would be unsuitable. For high-surface-area substrates such as highly porous inorganics for catalysis or microstructured materials for electronics, the use of conventional solvents is not appropriate. The surface tension and viscosity of common solvents, along with poor wetting of the pores, cause very slow and often incomplete impregnation of precursor solutions into porous supports. The low surface tension of $scCO_2$ not only permits better penetration and wetting of pores than liquid solvent but also avoids pore collapse that can occur on certain structures such as organic and silica aerogels with liquid solvents. This damage is caused during the drying step, as liquid solvents are removed from the substrate. Strong localized forces occur in the pores due to their small diameter in relation to the meniscus of retreating liquid. This causes cracking of the material. Even without this factor, extended heating of the substrate to remove liquid solvents can be costly and time consuming. In comparison, $scCO_2$ can be much more easily introduced and removed from such systems.

For polymer impregnation $scCO_2$ allows modification that would be difficult with conventional solvents. $scCO_2$ can be easily infused into polymers, without the need to dissolve the polymer and reblend, and displays a high permeation rate in virtually all polymers. The exposure of polymers to $scCO_2$ results in various extents of swelling and

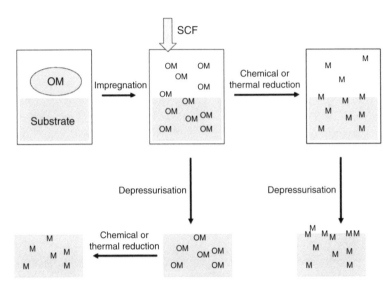

Figure 1.10. Supercritical fluid-mediated impregnation or deposition of nanoparticles into a substrate. OM represents a soluble organometallic precursor. Adapted from Reference [75]. Copyright 2006, with permission from Elsevier.

enhanced mobility of the polymer chains, which makes it possible to incorporate metallic precursors. The infusion of scCO$_2$ causes a lowering of the T_g of the polymer, allowing easier processability and molecular infusion. Moreover, the degree of polymer swelling, diffusion rates within the substrate, and the partitioning of precursors between the SCF and the swollen polymer can be controlled by density-mediated adjustments of solvent strength via changes in temperature and pressure [75, 76].

Regardless of the substrate used, the gas-like diffusivities of scCO$_2$ significantly enhance the kinetics of penetrant absorption, and no solvent residues are left after depressurization. The average particle size and size distribution of nanoparticles are affected by the precursor reduction method used, as well as the conditions, type, and amount of precursor in the system and the surface properties of the substrate.

The groundbreaking work in this field was undertaken by Watkins and McCarthy in 1995 and was the first use of an SCF for the impregnation of metal nanoparticles into a polymer substrate [76]. In this case the SCF used was scCO$_2$ and the polymers impregnated were poly(4-methyl-1-pentene) (PMP) and PTFE. The organometallic precursor used was dimethyl(cyclooctadiene)platinum(II) (CODPtMe$_2$), which was chosen because it was a readily available CVD precursor with good potential for scCO$_2$ solubility and had previously been reduced effectively with hydrogen to yield high-purity platinum films [77]. The precursor was infused into the polymers by a 4h impregnation time in scCO$_2$ at a temperature of 80 °C and pressure of 15.5 MPa. The precursor was reduced to platinum nanoparticles in the PMP by hydrogen reduction in scCO$_2$, hydrogen reduction after depressurization, and thermal decomposition in the presence of scCO$_2$. In PTFE the precursor was reduced by hydrogen in the presence of scCO$_2$. The results

(a)   (b)   (c)

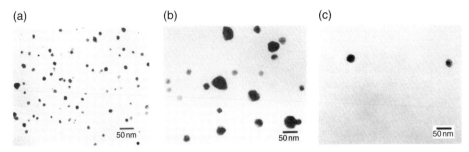

Figure 1.11. (a) Poly(4-methyl-1-pentene), hydrogen reduction after depressurization. (b) Poly (4-methyl-1-pentene), hydrogen reduction in the presence of scCO$_2$. (c) Poly(tetra-fluoroethylene), hydrogen reduction in the presence of scCO$_2$. Reproduced with permission from Reference [76]. Copyright 1995 American Chemical Society.

of these impregnation reactions provide a useful example to discuss the effect of changes in the system and can be observed from transmission electron micrographs of the nano-composites (Fig. 1.11).

The comparison of Figure 1.11a and b is useful for understanding the implications of the two alternative reduction routes shown in Figure 12, that is, reduction after or before depressurization. When the precursor in PMP was reduced after depressurization of the scCO$_2$ (Fig. 1.11a), the product appeared to be homogeneous throughout with nanoparticles of ~15 nm evenly distributed, apart from occasional larger clusters. When the precursor was reduced by hydrogen before venting the scCO$_2$, a sharp concentration gradient was seen close to the surface. The nanoparticles in the bulk of the PMP are mostly up to ~50 nm. However, close to the surface, which had a silver sheen, there was a greatly increased concentration of nanoparticles with diameters of ~50–100 nm. This difference is caused by the effect of reducing the metal while scCO$_2$ is still present in the polymer. The polymer is still scCO$_2$ swollen, and the precursor molecules are still able to diffuse freely during the reduction step. This allows the nanoparticles in the bulk to grow larger in diameter and also caused a greater deposition of metal into the surface layer. This is because the precursor is still able to infuse from the solvent phase into the polymer and because metal nanoparticles growing in the solvent phase will become unstable and deposit into the polymer surface. Therefore, when control of nanoparticle size and uniformity is desired, post-depressurization reduction is preferable.

We should also compare the difference in nanocomposite structure produced depending on the change of host polymer. Poly(tetrafluoroethylene) was impregnated with precursor that was reduced in the presence of scCO$_2$ (Fig. 1.11c) in exactly the same method as the PMP (Fig. 1.11b). A similar surface effect was noticed for the PTFE, compared to the PMP, in terms of there being more and larger nanoparticles at the polymer surface due to reduction occurring concurrently with impregnation. However, in the case of PTFE, the nanoparticles in the bulk of the polymer were consid-erably fewer in number and smaller in size than those produced in PMP in the same conditions. This demonstrates that although scCO$_2$ can be used to impregnate even very solvent-resistant polymers such as PTFE, the degree of nanoparticle incorporation is understandably affected by the uptake of precursor and this is determined by the solubility of the precursor in each given polymer and how easily the precursor can be infused into the polymer.

This method was expanded by Watkins' group to deposit different metal nanoparticles or films, including platinum, palladium, gold, rhodium, cobalt, nickel, and copper in scCO₂ by the reduction of desired precursors with hydrogen or alcohols [78–83]. Other groups then took this new technique forward, adapting it to other substrates and metals. Ye et al. prepared highly dispersed nanoparticles of palladium, rhodium, and ruthenium through reduction of adsorbed metal-β-diketone precursors by hydrogen in scCO₂ [84–87]. The nanoparticles were deposited onto functionalized multi-walled carbon nanotubes or SiO₂ nanotubes for catalytic applications. Preliminary experiments demonstrated that the supported Pd nanoparticles were effective catalysts for hydrogenation of olefins in carbon dioxide showing high electrocatalytic activity for oxygen reduction reaction indicative of potential fuel cell applications.

Saquing et al. demonstrated that metal films and highly dispersed nanoparticles could be produced simultaneously on the surface and within the matrix of porous monolithic materials by depositing platinum on carbon aerogel monoliths [88]. Further research led by Saquing and Erkey's group continued this work to other substrates including carbon black, silica aerogel, silica, gamma-alumina, and Nafion® 112 film [89]. Nafion® is a sulfonated tetrafluoroethylene copolymer developed by DuPont. The porous substrates and Nafion® film were impregnated with CODPtMe₂, which was reduced to elemental platinum by heat treatment in the presence of nitrogen gas. A comparison of nanocomposites produced under different conditions showed that both the metal contents and the particle sizes are controllable. Membrane–electrode assemblies of Nafion® membranes with incorporated palladium nanoparticles were prepared by this method and evaluated in direct methanol fuel cells. The Pd-impregnated Nafion® membranes showed reduced methanol crossover and gave higher cell performance than that of pure Nafion® membrane.

Substantial investigations into this field have been undertaken by the Howdle group at the University of Nottingham. The method used by the Howdle group has been generally similar to that commonly used, except instead of reducing the metallic precursor at ambient pressure following SCF impregnation, chemical reduction with pure hydrogen at an elevated pressure was used to synthesize silver nanoparticles in porous poly(styrene-divinylbenzene) beads and silica aerogels and palladium nanoparticles in silica aerogel [90, 91]. The palladium nanoparticles in silica aerogel supports were shown to demonstrate good catalytic activity for hydrogenation of cyclohexene in a continuous flow reactor. Similar impregnations allowed the clean synthesis of silver nanoparticles in polymer substrates for biomedical applications, and some success was demonstrated in initial applications testing [92, 93]. More recently the group was able to extend the versatility of this technique to the synthesis of gold nanoparticles supported on silica, polyamide, polypropylene, and PTFE [94]. Two precursors, dimethylacetylacetonate gold(III) and its fluorinated analogue dimethylhexafluoroacetylacetonate gold(III), were compared for their effectiveness in the production of the gold nanoparticles, which have catalytic applications. Interestingly the non-fluorinated organometallic precursor was found to produce more effectively and higher loaded nanocomposite material.

Some of the most advanced polymer results are evidenced by the researches of Boggess et al. and of Yoda et al. Boggess et al. performed the synthesis of highly reflective silver–polyimide films by impregnation of (1,5-cyclooctadiene)(1,1,1,5,5,5-hexafluoroacetylacetonate)silver(I) into polyimides [95]. Similarly, Yoda et al. synthesized Pt- and Pd-nanoparticle-dispersed polyimide films as a precursor for the preparation of metal-doped carbon molecular sieve membranes for hydrogen separation [96]. The impregnated precursors were Pt(II) acetylacetonate and Pd(II) acetylacetonate. It was found that the particle size increased with the increase of the impregnation temperature and impregnation time. The Pd-doped carbon molecular sieve membrane was successfully prepared by carbonation of the Pd-doped film and showed good hydrogen separation performance [97]. Yoda et al. also prepared platinum/silica aerogel nanocomposites by dissolving platinum(II) acetylacetonate in $scCO_2$ and impregnating into the silica gel during the supercritical drying stage [98].

These initial investigations into the use of $scCO_2$, though interesting and successful, only represent the beginning of research that will bring this exciting field into full understanding and effectiveness, but there is clearly a great potential for creating hybrid materials with improved properties.

The first section of this chapter has outlined the importance of the emergent field of nanocomposite functional materials and that the development of new routes to metal–polymer nanocomposites is a relevant and necessary area of research. Supercritical fluid technology has been explained as a useful technique with many advantages over conventional solvents.

This second half of the chapter gives some specific examples of how $scCO_2$ has been used to prepare metal nanoparticle-loaded polymer materials. The synthetic routes are outlined and the applications of these materials are discussed. These examples have been chosen to indicate the wide scope such materials may find use in, with fields covering optical/photonic sensors, antimicrobial medical materials, gas uptake, and catalysis.

1.16 SYNTHETIC METHOD FOR IMPREGNATING POLYMER FILMS WITH METAL NANOPARTICLES

The main criteria for choosing metal precursor complexes include solubility in $scCO_2$, affinity to polymer host, and decomposition temperature. The most commonly used silver precursor, for example, is (1,5-cyclooctadiene)(1,1,1,5,5,5-hexafluoroacetylacetonate)silver(I), Ag(hfac)-(1,5-COD), although many others have been used [99]. An ideal precursor should be $scCO_2$ soluble, be stable under high-pressure conditions, and degrade readily when required. The addition of fluorinated substituents has been shown to increase solubility in $scCO_2$ [55]. The addition of encapsulating ligands, such as multidentate chelates, completes the coordination sphere of the metal and essentially shields the metal from the $scCO_2$ [100]. The following impregnation procedure is representative of the method in general, but we will consider Ag(hfac)-(1,5-COD) as the example

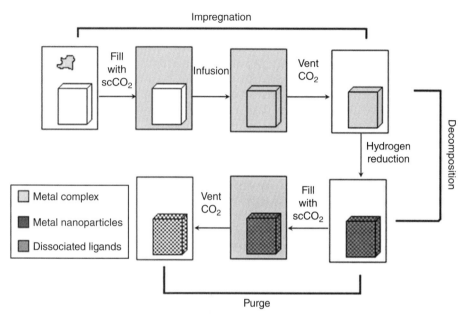

Figure 1.12. Schematic of a typical impregnation process as used by Howdle et al. consisting of three phases: impregnation, decomposition, and purging.

precursor. The impregnation process generally consists of three phases: impregnation, decomposition, and purge (Fig. 1.12). The first phase impregnates the precursor complex into the substrate. The precursor and substrate are placed in the autoclave, and the equipment is assembled before CO_2 is pumped in at a temperature just higher than the critical temperature of CO_2, for example, 40 °C. This ensures supercritical conditions while minimizing premature precursor decomposition. After leaving time for the infusion of the precursor into the substrate to occur (normally 24 hr), the autoclave is depressurized to vent the CO_2, leaving the precursor trapped in the substrate. Clearly, for this to happen effectively, it is essential that the precursor partitions into the polymer substrate. The second phase decomposes the precursor and leads to the formation of nanoparticles. In this step, the autoclave is filled with H_2 (typically at ~7 MPa) and heated to 80 °C to instigate the decomposition of the Ag precursor complex. This is thought to occur by a hydrogenolysis reaction resulting in Ag^0 and free dissociated ligands. The third phase is to repressurize the autoclave with $scCO_2$ at 40 °C, allowing the fluid to flow slowly through the autoclave for 24 hr to ensure complete extraction of the dissociated ligands. It is worth noting that this purging step is essential for processes in which by-products are undesirable, such as for antimicrobial materials that may be implanted in patients. However, for other materials, a negligible concentration of dissociated ligands might not be problematic, and this step can be omitted or shortened for expediency. This impregnation system has been used by Howdle et al. at the University of Nottingham to produce metal–polymer films for photonic [101, 102], catalytic [94],

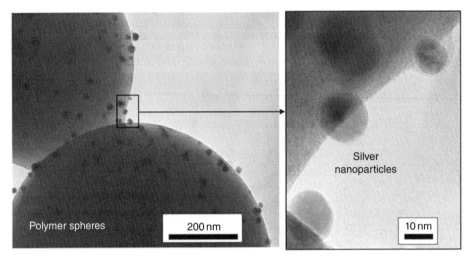

<u>Figure 1.13.</u> Bright-field TEM images detailing the surface location of the silver nanoparticles on PMMA microspheres following one-pot synthesis of silver–polymer nanocomposite in scCO2 in the presence of RAFT agent [103].

and medical applications [91, 92]. Following nanocomposite synthesis, there is always a need to characterize the particle morphology, size, distribution, and crystal structure of the metal phase inside the polymer matrix. This is particularly important when the properties of the metal nanoparticles are compared with their optical and biomedical performance. Among the characterization methods, transmission electron microscopy (TEM) is probably the most useful technique to directly probe the nanocomposites. Since the bulk polymer substrates tend to be too thick to transmit the electron beam, it is often necessary to microtome into transparent thin sections.

One striking alternative to the said process is a supercritical methodology that synthesizes not only the silver nanoparticles but also the polymer support in a simple process in $scCO_2$. Using reversible addition–fragmentation chain transfer (RAFT) polymerization agents, it is possible to produce silver nanoparticle-decorated polymer microspheres (Fig. 1.13) where the entire reaction takes place in a single step, in one pot, and in the absence of any conventional solvent [103]. In these reactions the polymer initiator is incorporated into the polymer surfactant, which is designed to also act as a stabilizing agent for the formation of metal nanoparticles, thus directing the nanoparticles onto the surface of the polymer spheres where they are ideally placed for many applications.

1.17 SILVER–POLYMER NANOCOMPOSITE FILMS FOR OPTICAL/PHOTONIC APPLICATIONS

Hasell et al. reported the clean impregnation of silver nanoparticles into a prefabricated polycarbonate strip in $scCO_2$, for application as an SERS device [101]. This method, even though producing nanoparticles inside the substrate *in situ*, did not cause any loss

(a)   (b)

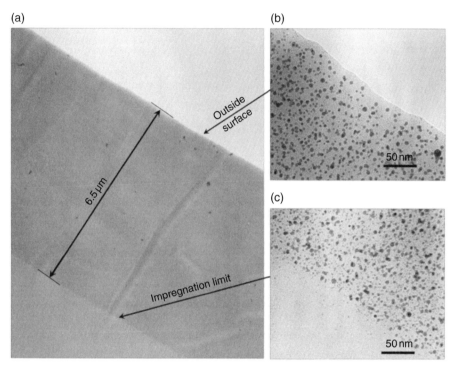

(c)

Figure 1.14. Bright-field TEM micrographs of a silver–polycarbonate nanocomposite sample prepared by cutting a 100 nm cross section of the material with an ultramicrotome fitted with a diamond blade. (a) The surface-located band of silver nanoparticles in the polycarbonate substrate. (b) A magnified section showing nanoparticles at the outside edge of the polymer. (c) A magnified section showing nanoparticles at the limit of furthest impregnation of the nanoparticles [101].

of transparency in the optical grade polymer. Transmission electron microscopy images of cross sections of the polymeric nanocomposite strips produced show clearly the silver nanoparticle distribution (Fig. 1.14). It is interesting to observe that there is a nanoparticle band of uniform thickness, of 6.5 µm, which runs along the outermost edge of the cross section of the polycarbonate substrate. There is a definite boundary of furthest impregnation. This was found to penetrate deeper, in a controllable manner, as a function of impregnation time. Discrete nanoparticles can be seen in the deposition band at the outermost surface of the polycarbonate, and they are roughly spherical in shape with an average diameter of 4.1 nm and a narrow size distribution. Such a silver–polycarbonate composite could allow the fabrication of optical and plasmonic devices that might exploit the mechanical robustness and flexibility of the polymers while providing protection of the particles from the surrounding environment (e.g., oxidation) to yield a high degree of temporal stability. The ability to precisely tune the impregnation depth of the nanoparticles on the micrometer scale gives this method an advantageous level of control for many potential applications.

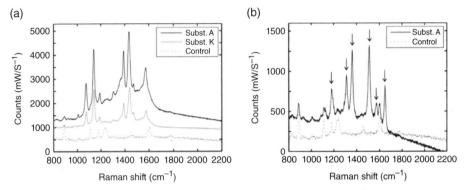

Figure 1.15. (a) Surface-enhanced Raman spectroscopy spectra of 4-ATP on the two metal–polycarbonate substrates (A and K) and (b) SERS spectrum of R6G taken on substrate A, where the arrows indicate the vibrational peaks of the dye. Both figures are plotted together with a control spectrum taken on undoped polycarbonate, and the spectra are offset for clarity.

The SERS activity of this nanocomposite was successfully demonstrated using both 4-aminothiophenol (4-ATP) and water-soluble Rhodamine 6G (R6G) as target molecules. These analytes were chosen because they have very distinct Raman features and form an ordered self-assembled monolayer on coinage-metal surfaces. The measured SERS spectra of the silver–polycarbonate film, following treatment with 4-ATP ethanol solution, are shown in Figure 1.15a. In comparison to a control sample of silver-free polycarbonate subject to identical treatment, the silver nanoparticle-loaded sample resulted in an enhancement factor greater than 10^7 of the Raman signal from the target molecule. Also, in the case of differently loaded samples, higher nanoparticle densities were found to lead to stronger SERS signals. A related investigation using R6G (Fig. 1.15b) suggests the overall SERS performance is also dependent on the bonding pattern of the target molecules to the silver.

Using the same silver–polycarbonate film substrate, Lagonigro et al. studied the time and spectrally resolved fluorescence [102]. Photoluminescence (PL) plays a crucial role in many aspects of biological and medical research with applications such as DNA sequencing [104], cell imaging [105], and sensing [106]. Large electric fields are generated by the plasmon modes of metal nanoparticles, and these can enhance photon–matter interaction to modify the excitation cross section [107] and the radiative decay rate. This phenomenon results in increased fluorescence intensities for molecules in close proximity to the metal [108]. Metal-enhanced fluorescence has led to the demonstration of optical probes with enhanced brightness, and these may play an important role in the next generation of photonic devices. This MEF activity further demonstrates the possible optical applications of silver nanostructures within polymer hosts. Metal-enhanced fluorescence measurement was conducted after a weak solution of coumarin 102, which served as a dye molecule, was spin coated onto the nanocomposite film substrate. The fluorescence spectra of coumarin 102, with time and spectral resolution,

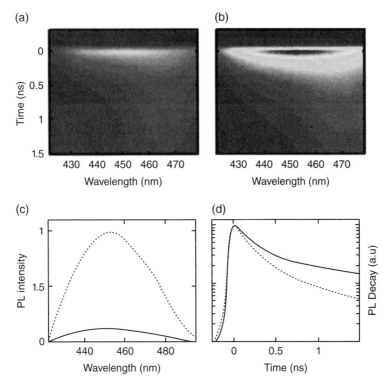

Figure 1.16. Time and spectrally resolved PL for coumarin on (a) silver-free polycarbonate and (b) silver–polycarbonate nanocomposite film. (c) Integrated PL spectra and (d) average PL decays on polycarbonate (solid) and silver-loaded polycarbonate (dotted). Reprinted with permission from Reference [102]. Copyright 2008, American Institute of Physics.

of both the silver-impregnated polycarbonate and silver-free polycarbonate are shown in Figure 1.16. The silver nanoparticles inside the polycarbonate give an enhancement of ~8.5 times the emission intensity, accompanied by a reduction in the photon lifetime. These preliminary investigations show how the excellent plasmonic properties of silver nanoparticles can also be utilized to yield large electric fields in the vicinity of fluorophores. In order for MEF to be conveniently used in routine procedures, the substrates need to be robust, stable, and, most importantly, biocompatible. Polycarbonate is known to have excellent biocompatibility, as well as optical and mechanical properties [30], and these composite structures offer a number of advantages over previously reported metal–polymer substrates fabricated via surface functionalization [109]. Owing to the versatility of this supercritical technique and its applicability to a range of prefabricated polymer films and devices and the abundance of suitable metal precursors, such substrates can be readily tailored for a wide range of applications in medicine and biology.

1.18 SILVER–POLYMER NANOCOMPOSITE FILMS FOR ANTIMICROBIAL APPLICATIONS

Due to the development of antibiotic resistance in many strains of infectious diseases, there is a need for alternative antibacterial agents. Metallic silver has been used throughout history for the treatment of wounds and burns. This antimicrobial property is related to the amount and rate of silver released as ions, in which state it can bind to proteins in the bacterial cell wall and nuclear membrane, leading to cell distortion and death. It is because of this effect that nanoparticlulate silver has been found to be so effective due to its large surface to volume ratio and crystallographic surface structure. Polymers are extensively used for medical and healthcare products because of their low toxicity and high biocompatibility, making them ideal hosts for biochemically active inorganic species. Polymeric substrates therefore offer a sustainable support to silver nanoparticles and allow such composites to be fabricated into devices. The bactericidal properties of silver nanoparticles have been shown to be size dependent; smaller nanoparticles, and especially those with diameters in the range of ~1–10 nm, have been shown to have the highest activity [110]. Supercritical processing is highly attractive as a route to produce clean silver–polymer nanocomposite materials for biomedical applications, leading Furno et al. to use $scCO_2$ for the impregnation of silver nanoparticles into a silicone support, which showed clear antimicrobial activity [93]. In order to determine the biomedical performance of the silver–polymer nanocomposites, the researchers investigated the release of silver ions, the adherence of silver to the substrate surface, and the killing of bacteria. The results of these tests suggest the nanocomposites have a clear persisting and diffusible activity to kill bacteria. Morley et al. later reported silver nanoparticle deposition into biodegradable ultrahigh-molecular-weight polyethylene (UHMWPE) using $scCO_2$ [92]. Ultrahigh-molecular-weight polyethylene is known as an extremely difficult substrate to process, which significantly demonstrates the ability of SCF processing to introduce metal nanoparticles into structures not accessible by conventional means. By *in situ* impregnation with $scCO_2$, nanocomposites with enhanced biocompatibility and antimicrobial activity were prepared. More recently, the same $scCO_2$ processing method was used to introduce silver nanoparticles into catheter tubing made of polydimethylsiloxane (PDMS) polymer. In this case an extended $scCO_2$ purge step was employed to ensure that all of the dissociated ligands were removed post processing [99]. The silver–PDMS nanocomposites produced were analyzed for their antimicrobial efficacy against a range of pathogens including the Gram-positive bacteria MRSA, the Gram-negative bacteria *E. coli*, and the yeast *C. albicans*. These silver–polymer composites showed significant activity towards both Gram-positive and Gram-negative adhered bacteria, eradicating 100% of adhered bacteria within the 72 hour time frame of the test. Furthermore, these tests demonstrated the viability of thorough CO_2 purging by showing there was no detectable residue of decomposed ligands left in the nanocomposites, as determined by gas chromatography with a detection limit of 7 ppm. This is particularly important since dissociated ligands from the precursor molecules used, such as the Ag(hfac)(1,5-COD) used in this instance, could have toxicological effects on surrounding tissues if implanted or indeed other unwanted effects in alternative medical or otherwise sensitive applications.

1.19 PALLADIUM–POLYMER NANOCOMPOSITE FILMS FOR CATALYSIS OR HYDROGEN UPTAKE APPLICATIONS

A research group at the University of Liverpool has demonstrated that the same simple impregnation strategy can also be used to introduce metal nanoparticles into the porous polymeric systems [111]. Microporous polymeric materials are receiving much current interest because of their potential applications in areas such as gas sorption, separation, and heterogeneous catalysis. Palladium nanoparticles were dispersed uniformly throughout a microporous poly(aryleneethynylene) material by the infusion of a CO_2-soluble Pd precursor, palladium(II) hexafluoroacetylacetonate. The resulting composite showed good thermal stability and has potential applications in gas storage and heterogeneous catalysis. The porosity of the polymeric support was maintained after inclusion of the metal, and the hydrogen uptake at room temperature was increased with respect to the unloaded porous support. Hydrogen storage by spillover has been suggested as a promising approach to enhancing the ambient-temperature hydrogen storage capacity of nanostructured materials. Conceptually, spillover involves the dissociation of hydrogen molecules on metal nanoparticles followed by subsequent hydrogen atom migration onto the nearby substrate by surface diffusion. Palladium can both chemisorb atomic hydrogen and also dissolve it to form bulk hydride phases and is capable of adsorbing hydrogen dissociatively and thereby act as a source of hydrogen atoms to migrate onto the substrate surface. Supported Pd nanoparticles are also relevant as catalysts for organic and industrial reactions. It is important for many of these applications that the nanoparticles should be small in size and well dispersed in a robust support of high surface area. It has been demonstrated that conjugated microporous polymer (CMP) systems are able to function as effective support and stabilization substrates for active metal nanoparticles, such as palladium [112]. The target substrate used for the scCO$_2$ deposition of Pd nanoparticles was based on a CMP [113] material. These CMP networks have a tunable micropore size distribution and surface area that is controlled by the length of the rigid organic linkers. Conjugated microporous polymers are relatively thermally robust and chemically stable. In this case, the impregnation of a conjugated poly(aryleneethynylene) polymer with palladium nanoparticles was achieved using the same general strategy described earlier and a relatively simple experimental setup (Fig. 1.17). The use of hydrogen for the reduction step was omitted as it was found that simple heating, to over 300 °C, could be more simply used to form nanoparticles as the polymer was stable to these high temperatures.

The results found for the impregnation step in this research provide an ideal example to discuss what is known as the partitioning effect. In order to introduce an SCF-soluble species into an insoluble substrate, the highest pressure (and therefore highest density and solubility) is very rarely found to be the optimal condition. The mass increase of the polymer after impregnation demonstrates the efficiency of the loading (Fig. 1.18). The optimum pressure was found to be ~8 MPa, just above the critical pressure of CO_2 (7.38 MPa). This can be explained by two competing factors: solubility and solid–fluid partitioning. At lower pressures, the Pd complex is not soluble enough to be effectively transported into the polymer, whereas at higher pressures, the solubility increases to the point that the Pd complex preferentially partitions into the CO_2 outside of the polymer.

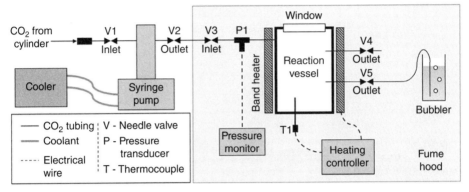

Figure 1.17. Schematic representation of the equipment used for the supercritical impregnation. The reaction vessel consists of a $10\,cm^3$ volume stainless steel autoclave fitted with a sapphire view window. Pressure is controlled by the syringe pump, heating by an external heater with internal thermocouple, and venting speed by using an outlet bubbler [111].

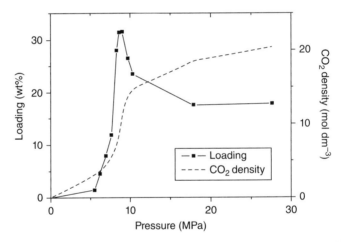

Figure 1.18. Loading of Pd complex as a function of CO_2 pressure at 40 °C [111].

Investigation of the products by scanning and TEM (Fig. 1.19) revealed larger nanoparticles on the surface of the polymer with smaller nanoparticles inside the polymer network. The ~5–10 nm surface-bound nanoparticles were ascribed to Pd complex that was deposited on the surface during depressurization. These particles were able to grow more freely and hence agglomerate, whereas the growth of particles formed inside the microporous network was restricted by the pore dimensions. The nanoparticles inside the network were found to be limited to ~1–3 nm in diameter, in close agreement with the average diameter of the pores of the polymer. The large number of smaller nanoparticles as well as their dispersed nature throughout the pores of the substrate would be difficult to achieve with conventional impregnation techniques, such as solvent infusion or CVD.

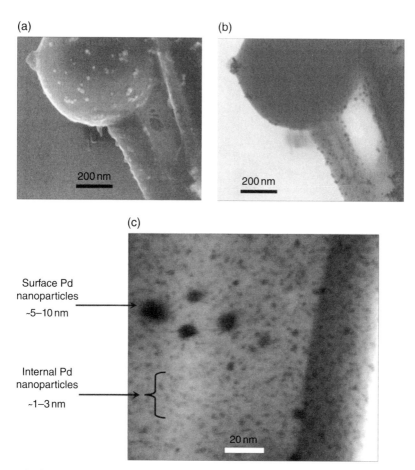

Figure 1.19. Electron microscopy of a Pd-loaded CMP sample. (**a**) Scanning mode image (left) and (**b**) transmission mode (right) of the same area of microporous polymer. At this magnification, only larger nanoparticles located on the surface are visible. (**c**) Under higher-magnification transmission mode, a bimodal nanoparticle size distribution is observed where the smaller, internally located nanoparticles can be observed [111].

Smaller nanoparticles and extensive dispersion are in principle desirable for increased hydrogen uptake. Gas sorption measurements of these polymers indicated that porosity was maintained after impregnation of Pd nanoparticles and that hydrogen uptake, at room temperature and low pressure, was significantly enhanced by the addition of the nanoparticles (Fig. 1.20). This research showed that in addition to conventional polymers, scCO$_2$ is also able to provide a simple and effective processing route to generate metal nanoparticles *in situ* within prefabricated CMPs. The nanoparticles are well dispersed throughout the material and show excellent size control within the pore matrix. Such polymer–metal composite materials may have future advantages over materials such as carbon because they can couple greater synthetic versatility with

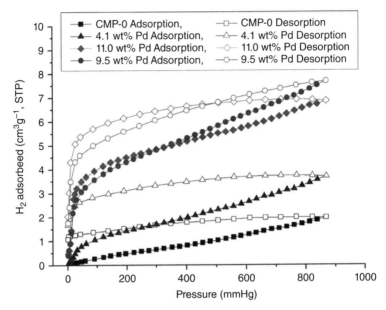

Figure 1.20. Low-pressure H_2 isotherms at 20 °C for a series of microporous polymer samples. The hydrogen uptake is significantly increased by the inclusion of Pd nanoparticles.

adequate thermal and chemical robustness. Such systems may have potential as substrates for hydrogen spillover—for example, in catalytic processes—but the amounts of H_2 adsorbed at ambient temperature are still below the levels that are interesting for practical gas storage applications.

1.20 CONCLUSIONS

This chapter has outlined the importance of the emergent field of nanocomposite functional materials and has shown that the development of new routes to metal–polymer nanocomposites is a relevant and necessary area of study. An introduction and overview into the fields of nanoparticles, polymers, and SCFs has been provided, with the intention of giving readers with no previous knowledge of these fields sufficient background knowledge.

The work reported in this chapter demonstrates the possibility of using $scCO_2$ techniques for the formation of optical quality polymer–metal nanocomposite material. It has been shown that $scCO_2$ is a viable technique to produce uniform and transparent plasmonic nanocomposites. The polymeric substrate (polycarbonate) is regularly used to construct optical devices such as lenses and windows. As the nanoparticles are introduced *in situ* and post fabrication of the material, without loss of transparency, it is easy to see the possible benefits of using such a system to introduce metal particles into devices after they have been processed into the final form. This neatly avoids any aggregation or degradation of the uniform nanoparticle distribution that would occur during the fabrication of polymer feedstock into a desired architecture. The impregnation process shows good potential for control of both depth and size/concentration of the nanoparticles by modification of time and pressure variables. The products have been demonstrated as excellent SERS and MEF enhancement materials, which could find application in the next generation of plasmonic devices and sensors and in particular biological or in vivo monitoring. These substrates offer a number of significant benefits in that they are cheap, flexible, mechanically robust, and temporally stable so that they can be stored for long lengths of time, easily handled and discarded after use to avoid cross contamination. There are a wide range of optical and biomedical potential applications, and a key conclusion is that $scCO_2$ provides a very flexible and clean route to novel materials and properties.

It has been illustrated that as well as impregnating preformed solid polymers with metal nanoparticles, it is also possible to directly synthesis the polymer itself at the same time as the metal nanoparticles, all taking place in carbon dioxide. This was achieved by the use of an appropriately designed RAFT/stabilizing agent that allows simple and effective synthesis of polymer microspheres decorated with silver nanoparticles. The surface location of the nanoparticles gives the polymer powder excellent potential for catalytic, antibacterial, and biosensing applications. This synthetic route has the additional advantages of being a single step, one-pot reaction that does not require conventional solvents at any stage during the process. Furthermore, the versatility of this technique suggests its applicability for a wide range of polymer nanoparticle composite materials.

Supercritical CO_2 has also been shown to provide a simple and effective processing route to generate metal nanoparticle-loaded CMPs. The nanoparticles are again well dispersed throughout the material and show excellent size control within the pore matrix.

There are certainly limitations to the use of $scCO_2$, for example, in the solubility of reagents and the necessity of using high pressures, and it certainly does not provide a universal solution to making nanocomposites. However, there are also clear benefits and specific examples where materials can be fabricated using $scCO_2$ that simply could not be accessed using conventional technologies. Whatever routes are chosen, it is clear that by improving the quality of the nanocomposites and understanding the mechanism of their optical and biomedical activities, useful devices for practical applications will result.

REFERENCES

[1] L. Nicolais and G. Carotenuto, *Metal-Polymer Nanocomposites*, Wiley, Hoboken, 2005.
[2] E. Roduner, *Chemical Society Reviews*, 2006, **35**, 583–592.
[3] C. N. R. Rao, P. J. Thomas and G. U. Kulkarni, *Nanocrystals: Synthesis, Properties and Applications*, Springer, Berlin-Heidelberg-New York, 2007.
[4] F. Raimondi, G. G. Scherer, R. Kotz and A. Wokaun, *Angewandte Chemie-International Edition*, 2005, **44**, 2190–2209.
[5] C. R. Henry, *Surface Science Reports*, 1998, **31**, 231–325.
[6] A. Haruta, *Chemical Record*, 2003, **3**, 75–87.
[7] M. Haruta, *Cattech*, 2002, **6**, 102–115.

[8] M. Haruta and M. Date, *Applied Catalysis a-General*, 2001, **222**, 427–437.

[9] M. C. Daniel and D. Astruc, *Chemical Reviews*, 2004, **104**, 293–346.

[10] F. Boccuzzi, A. Chiorino, M. Manzoli, P. Lu, T. Akita, S. Ichikawa and M. Haruta, *Journal of Catalysis*, 2001, **202**, 256–267.

[11] M. Mavrikakis, P. Stoltze and J. K. Norskov, *Catalysis Letters*, 2000, **64**, 101–106.

[12] P. N. Prasad, *Current Opinion In Solid State & Materials Science*, 2004, **8**, 11–19.

[13] D. J. Norris and M. G. Bawendi, *Physical Review B*, 1996, **53**, 16338–16346.

[14] G. Schmid, *Nanoparticles: From Theory to Applications*, Wiley, New York, 2004.

[15] C. F. Bohren and D. R. Huffman, *Absorption and Scattering of Light by Small Particles*, Wiley, New York, 1983.

[16] P. N. Prasad, *Nanophotonics*, Wiley, Hoboken, New Jersey, 2004.

[17] K. L. Kelly, E. Coronado, L. L. Zhao and G. C. Schatz, *Journal of Physical Chemistry B*, 2003, **107**, 668–677.

[18] N. K. Grady, N. J. Halas and P. Nordlander, *Chemical Physics Letters*, 2004, **399**, 167–171.

[19] S. Link and M. A. El-Sayed, *Journal of Physical Chemistry B*, 1999, **103**, 8410–8426.

[20] D. D. Evanoff and G. Chumanov, *Chemphyschem*, 2005, **6**, 1221–1231.

[21] Z. X. Liu, H. H. Wang, H. Li and X. M. Wang, *Applied Physics Letters*, 1998, **72**, 1823–1825.

[22] S. Shi, W. Ji, S. H. Tang, J. P. Lang and X. Q. Xin, *Journal of American Chemical Society*, 1994, **116**, 3615–3616.

[23] Y. P. Sun, J. E. Riggs, H. W. Rollins and R. Guduru, *Journal of Physical Chemistry B*, 1999, **103**, 77–82.

[24] L. Francois, M. Mostafavi, J. Belloni, J. F. Delouis, J. Delaire and P. Feneyrou, *Journal of Physical Chemistry B*, 2000, **104**, 6133–6137.

[25] R. B. Martin, M. J. Meziani, P. Pathak, J. E. Riggs, D. E. Cook, S. Perera and Y. P. Sun, *Optical Materials*, 2007, **29**, 788–793.

[26] V. Liberman, M. Rothschild, O. M. Bakr and F. Stellacci, *Journal of Optics*, 2010, **12**.

[27] S. Porel, N. Venkatrarn, D. N. Rao and T. P. Radhakrishnan, *Journal of Applied Physics*, 2007, **102**.

[28] E. Smith and G. Dent, *Modern Raman Spectroscopy - A Practical Approach*, Wiley, Chichester, 2005.

[29] K. Kneipp, Y. Wang, H. Kneipp, L. T. Perelman, I. Itzkan, R. Dasari and M. S. Feld, *Physical Review Letters*, 1997, **78**, 1667–1670.

[30] K. Aslan, P. Holley and C. D. Geddes, *Journal Of Materials Chemistry*, 2006, **16**, 2846–2852.

[31] K. Aslan, I. Gryczynski, J. Malicka, E. Matveeva, J. R. Lakowicz and C. D. Geddes, *Current Opinion in Biotechnology*, 2005, **16**, 55–62.

[32] J. P. Abid, A. W. Wark, P. F. Brevet and H. H. Girault, *Chemical Communications*, 2002, 792–793.

[33] H. H. Huang, X. P. Ni, G. L. Loy, C. H. Chew, K. L. Tan, F. C. Loh, J. F. Deng and G. Q. Xu, *Langmuir*, 1996, **12**, 909–912.

[34] S. A. Dong and S. P. Zhou, *Materials Science and Engineering B-Solid State Materials for Advanced Technology*, 2007, **140**, 153–159.

[35] H. Hirai, Y. Nakao and N. Toshima, *Journal of Macromolecular Science-Chemistry*, 1979, **A13**, 727–750.

[36] M. Brust, M. Walker, D. Bethell, D. J. Schiffrin and R. Whyman, *Journal of the Chemical Society-Chemical Communications*, 1994, 801–802.

[37] J. A. Creighton, C. G. Blatchford and M. G. Albrecht, *Journal Of The Chemical Society-Faraday Transactions Ii*, 1979, **75**, 790–798.

[38] P. C. Lee and D. Meisel, *Journal Of Physical Chemistry*, 1982, **86**, 3391–3395.

[39] H. Bonnemann and R. M. Richards, *European Journal Of Inorganic Chemistry*, 2001, 2455–2480.

[40] M. V. Volpe, A. Longo, L. Pasquini, V. Casuscelli and G. Carotenuto, *Journal Of Materials Science Letters*, 2003, **22**, 1697–1699.

[41] G. Carotenuto and L. Nicolais, *Journal of Materials Chemistry*, 2003, **13**, 1038–1041.

[42] E. Hutter and J. H. Fendler, *Advanced Materials*, 2004, **16**, 1685–1706.

[43] G. Carotenuto, B. Martorana, P. B. Perlo and L. Nicolais, *Journal of Material Chemistry*, 2003, **13**, 2927–2930.

[44] S. Matsuda and S. Ando, *Japanese Journal of Applied Physics Part 1-Regular Papers Short Notes & Review Papers*, 2005, **44**, 187–192.

[45] S. Porel, S. Singh, S. S. Harsha, D. N. Rao and T. P. Radhakrishnan, *Chemical Matters*, 2005, **17**, 9–12.

[46] S. Matsuda, S. Ando and T. Sawada, *Electron. Lett.*, 2001, **37**, 706–707.

[47] A. V. Gaikwad and T. K. Rout, *Journal of Materials Chemistry*, 2011, **21**, 1234–1239.

[48] S. Koizumi, S. Matsuda and S. Ando, *Journal of Photopolymer Science and Technology*, 2002, **15**, 231–236.

[49] A. L. Stepanov, S. N. Abdullin, V. Y. Petukhov, Y. N. Osin, R. I. Khaibullin and I. B. Khaibullin, *Philosophical Magazine B*, 2000, **80**, 23–28.

[50] A. L. Stepanov, S. N. Abdullin and I. B. Khaibullin, *Journal of Non-Crystalline Solids*, 1998, **223**, 250–253.

[51] M. Okumura, S. Nakamura, S. Tsubota, T. Nakamura, M. Azuma and M. Haruta, *Catalysis Letters*, 1998, **51**, 53–58.

[52] T. Clifford, *Fundamentals of Supercritical Fluid*. OUP, Oxford, 1999.

[53] T. Andrews, *Philosophical Magazine*, 1870, 150–153.

[54] J. B. Hannay and J. Hogarth, *Proceedings of Royal Society B London*, 1879, **29**, 324–326.

[55] J. A. Darr and M. Poliakoff, *Chemical Reviews*, 1999, **99**, 495–541.

[56] A. I. Cooper, *Journal of Material Chemistry*, 2000, **10**, 207–234.

[57] F. Furno, P. Licence, S. M. Howdle and M. Poliakoff, *Actualite Chimique*, 2003, 62–66.

[58] J. R. Hyde, P. Licence, D. Carter and M. Poliakoff, *Applied Catalysis a-General*, 2001, **222**, 119–131.

[59] E. J. Beckman, *Journal of Supercritical Fluids*, 2004, **28**, 121–191.

[60] C. F. Kirby and M. A. McHugh, *Chemical Reviews*, 1999, **99**, 565–602.

[61] D. L. Tomasko, H. B. Li, D. H. Liu, X. M. Han, M. J. Wingert, L. J. Lee and K. W. Koelling, *Industrial and Engineering Chemistry Research*, 2003, **42**, 6431–6456.

[62] J. L. Kendall, D. A. Canelas, J. L. Young and J. DeSimone, *Chemical Reviews*, 1999, **99**, 543–563.

[63] W. Wang, A. Naylor and S. M. Howdle, *Macromolecules*, 2003, **36**, 5424–5427.

[64] C. D. Wood and A. I. Cooper, *Macromolecules*, 2001, **34**, 5–8.

[65] J. M. Desimone, Z. Guan and C. S. Elsbernd, *Science*, 1992, **257**, 945–947.

[66] J. M. Desimone, E. E. Maury, Y. Z. Menceloglu, J. B. McClain, T. J. Romack and J. R. Combes, *Science*, 1994, **265**, 356–359.

[67] P. Christian, M. R. Giles, R. M. T. Griffiths, D. J. Irvine, R. C. Major and S. M. Howdle, *Macromolecules*, 2000, **33**, 9222–9227.

[68] P. Christian, S. M. Howdle and D. J. Irvine, *Macromolecules*, 2000, **33**, 237–239.

[69] H. M. Woods, M. Silva, C. Nouvel, K. M. Shakesheff and S. M. Howdle, *Journal of Materials Chemistry*, 2004, **14**, 1663–1678.

[70] K. J. Thurecht, A. M. Gregory, S. Villarroya, J. X. Zhou, A. Heise and S. M. Howdle, *Chemical Communications*, 2006, 4383–4385.

[71] S. Villarroya, K. J. Thurecht, A. Heise and S. M. Howdle, *Chemical Communications*, 2007, 3805–3813.

[72] S. Villarroya, K. J. Thurecht and S. M. Howdle, *Green Chemistry*, 2008, **10**, 863–867.

[73] J. X. Zhou, S. Villarroya, W. X. Wang, M. F. Wyatt, C. J. Duxbury, K. J. Thurecht and S. M. Howdle, *Macromolecules*, 2007, **40**, 2276–2276.

[74] I. Kikic and F. Vecchione, *Current Opinion in Solid State & Materials Science*, 2003, **7**, 399–405.

[75] Y. Zhang and C. Erkey, *Journal of Supercritical Fluids*, 2006, **38**, 252–267.

[76] J. J. Watkins and T. J. McCarthy, *Chemistry of Materials*, 1995, **7**, 1991.

[77] R. Kumar, S. Roy, M. Rashidi and R. J. Puddephatt, *Polyhedron*, 1989, **8**, 551–553.

[78] Y. F. Zong and J. J. Watkins, *Chemistry of Materials*, 2005, **17**, 560–565.

[79] A. Cabanas, D. P. Long and J. J. Watkins, *Chemistry of Materials*, 2004, **16**, 2028–2033.

[80] E. T. Hunde and J. J. Watkins, *Chemistry of Materials*, 2004, **16**, 498–503.

[81] A. Cabanas, X. Y. Shan and J. J. Watkins, *Chemistry of Materials*, 2003, **15**, 2910–2916.

[82] A. Cabanas, J. M. Blackburn and J. J. Watkins, *Microelectronic Engineering*, 2002, **64**, 53–61.

[83] J. M. Blackburn, D. P. Long, A. Cabanas and J. J. Watkins, *Science*, 2001, **294**, 141–145.

[84] X. R. Ye, J. B. Talbot, S. H. Jin, Y. H. Lin and C. M. Wai, *Abstracts of Papers of the American Chemical Society*, 2005, **229**, U926–U926.

[85] X. R. Ye, Y. H. Lin, C. M. Wai, J. B. Talbot and S. H. Jin, *Journal of Nanoscience and Nanotechnology*, 2005, **5**, 964–969.

[86] X. R. Ye, H. F. Zhang, Y. H. Lin, L. S. Wang and C. M. Wai, *Journal of Nanoscience and Nanotechnology*, 2004, **4**, 82–85.

[87] X. R. Ye, Y. H. Lin, C. M. Wang, M. H. Engelhard, Y. Wang and C. M. Wai, *Journal of Materials Chemistry*, 2004, **14**, 908–913.

[88] C. D. Saquing, D. Kang, M. Aindow and C. Erkey, *Microporous and Mesoporous Materials*, 2005, **80**, 11–23.

[89] Y. Zhang, D. F. Kang, C. Saquing, M. Aindow and C. Erkey, *Industrial & Engineering Chemistry Research*, 2005, **44**, 4161–4164.

[90] K. S. Morley, P. Licence, P. C. Marr, J. R. Hyde, P. D. Brown, R. Mokaya, Y. D. Xia and S. M. Howdle, *Journal Of Materials Chemistry*, 2004, **14**, 1212–1217.

[91] K. S. Morley, P. C. Marr, P. B. Webb, A. R. Berry, F. J. Allison, G. Moldovan, P. D. Brown and S. M. Howdle, *Journal Of Materials Chemistry*, 2002, **12**, 1898–1905.

[92] K. S. Morley, P. B. Webb, N. V. Tokareva, A. P. Krasnov, V. K. Popov, J. Zhang, C. J. Roberts and S. M. Howdle, *European Polymer Journal*, 2007, **43**, 307–314.

[93] F. Furno, K. S. Morley, B. Wong, B. L. Sharp, P. L. Arnold, S. M. Howdle, R. Bayston, P. D. Brown, P. D. Winship and H. J. Reid, *Journal of Antimicrobial Chemotherapy*, 2004, **54**, 1019–1024.

[94] B. Wong, S. Yoda and S. M. Howdle, *Journal of Supercritical Fluids*, 2007, **42**, 282–287.

[95] R. K. Boggess, L. T. Taylor, D. M. Stoakley and A. K. StClair, *Journal of Applied Polymer Science*, 1997, **64**, 1309–1317.

[96] S. Yoda, A. Hasegawa, H. Suda, Y. Uchimaru, K. Haraya, T. Tsuji and K. Otake, *Chemistry Of Materials*, 2004, **16**, 2363–2368.

[97] H. Suda, S. Yoda, A. Hasegawa, T. Tsuji, K. Otake and K. Haraya, *Desalination*, 2006, **193**, 211–214.

[98] S. Yoda, Y. Takebayashi, T. Sugeta and K. Otake, *Journal Of Non-Crystalline Solids*, 2004, **350**, 320–325.

[99] J. X. Yang, T. Hasell, D. C. Smith and S. M. Howdle, *Journal of Materials Chemistry*, 2009, **19**, 8560–8570.

[100] K. Morley, *The clean preparation of nanocomposite materials: a supercritical route*, PhD Thesis, University of Nottingham, 2003.

[101] T. Hasell, L. Lagonigro, A. C. Peacock, S. Yoda, P. D. Brown, P. J. A. Sazio and S. M. Howdle, *Advanced Functional Materials*, 2008, **18**, 1265–1271.

[102] L. Lagonigro, A. C. Peacock, S. Rohrmoser, T. Hasell, S. M. Howdle, P. J. A. Sazio and P. G. Lagoudakis, *Applied Physics Letters*, 2008, **93**.

[103] T. Hasell, K. J. Thurecht, R. D. W. Jones, P. D. Brown and S. M. Howdle, *Chemical Communications*, 2007, 3933–3935.

[104] L. M. Smith, J. Z. Sanders, R. J. Kaiser, P. Hughes, C. Dodd, C. R. Connell, C. Heiner, S. B. H. Kent and L. E. Hood, *Nature*, 1986, **321**, 674–679.

[105] D. J. Stephens and V. J. Allan, *Science*, 2003, **300**, 82–86.

[106] P. K. Jain, X. Huang, I. H. El-Sayed and M. A. El-Sayad, *Plasmonics*, 2007, **2**, 107–118.

[107] J. S. Biteen, D. Pacifici, N. S. Lewis and H. A. Atwater, *Nano Letters*, 2005, **5**, 1768–1773.

[108] J. R. Lakowicz, *Analytical Biochemistry*, 2005, **337**, 171–194.

[109] K. Aslan, Z. Leonenko, J. R. Lakowicz and C. D. Geddes, *Journal of Fluorescence*, 2005, **15**, 643–654.

[110] M. Rai, A. Yadav and A. Gade, *Biotechnology Advances*, 2009, **27**, 76–83.

[111] T. Hasell, C. D. Wood, R. Clowes, J. T. A. Jones, Y. Z. Khimyak, D. J. Adams and A. I. Cooper, *Chem. Mat.*, 2010, **22**, 557–564.

[112] J. Schmidt, J. Weber, J. D. Epping, M. Antonietti and A. Thomas, *Advanced Materials*, 2009, **21**, 702.

[113] J. X. Jiang, F. Su, A. Trewin, C. D. Wood, N. L. Campbell, H. Niu, C. Dickinson, A. Y. Ganin, M. J. Rosseinsky, Y. Z. Khimyak and A. I. Cooper, *Angewandte Chemie-International Edition*, 2008, **47**, 1167–1167.

2

IN SITU SYNTHESIS OF POLYMER-EMBEDDED NANOSTRUCTURES

W. R. Caseri

Department of Materials, ETH Zürich, Zürich, Switzerland

2.1 INTRODUCTION

Embedding of inorganic nanoparticles in polymers has attracted increasing interest in the last two decades. Namely, inorganic nanoparticles can introduce specific physical properties into composites with polymers (examples are given later), and these properties can differ from those of analogous materials with larger particles. For instance, magnetic properties of nanoparticles can distinguish from those of their bulk materials counterparts [1, 2]; superparamagnetism can emerge for particles with diameters below the critical size of magnetic domain wall formation, provided that most particles are well separated from each other [3]. Also, optical properties of metal or semiconductor nanoparticles frequently change with particle size [4]; that is, colors, UV–visible (Vis) absorption spectra, or refractive indices may depend considerably on the size of metallic or semiconductor particles. A general optical property of nanoparticles is the pronounced reduction in light scattering when the particles' dimensions fall well below the wavelength of incident light. This effect renders nanocomposites transparent or translucent provided that the particles do not agglomerate. If the particles agglomerate, the materials appear opaque, unless the refractive indices of polymer and particles are equal. Therefore, considerable efforts have been carried out to prevent particle agglomeration.

Nanocomposites: In Situ *Synthesis of Polymer-Embedded Nanostructures*, First Edition.
Edited by Luigi Nicolais and Gianfranco Carotenuto.
© 2014 John Wiley & Sons, Inc. Published 2014 by John Wiley & Sons, Inc.

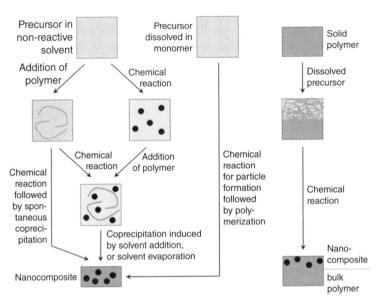

Figure 2.1. Schematic overview on methods used for embedding nanoparticles into polymers.

Agglomeration of inorganic particles is favored in physical terms by the high surface free energies of inorganic materials, which typically amount to 500–2000 mJ/m^2 compared to 20–50 mJ/m^2 for polymers [5–7]. Hence, surfaces of inorganic particles are thermodynamically in particularly unfavorable states, and therefore inorganic particles show a strong tendency to reduce their surface free energies. This may happen by coating the inorganic surfaces with organic molecules or by reduction of surface area via agglomeration. Such agglomerates, as typically present in isolated powders of uncoated nanoparticles, are difficult to break up into individual primary particles by convenient methods of nanocomposite preparation. In order to avoid particle agglomeration in nanocomposites, *in situ* synthesis of particles has been established as a successful route. This route also allows the preparation of materials with high volume fractions of nanoparticles, such as 50% v/v. We have been active in this area for 20 years and have dealt with representative compounds of essential classes of inorganic materials for *in situ* preparation of nanoparticles and their embedding into polymers. These classes are (i) metals, (ii) metal oxides, and (iii) metal sulfides. By means of corresponding examples from our laboratory, some principles of nanoparticle synthesis and nanocomposite preparation as well as some specific materials properties introduced by particles with nanoscale dimensions will be exposed in the following.

We synthesized the particles in liquid media and subsequently embedded them in polymers by application of different methods, as schematically illustrated in an overview in Figure 2.1. Mostly, the particles were synthesized either in the presence of a dissolved or swollen polymer or the polymer was dissolved afterwards in the particle dispersion. Note that a polymer can stabilize or destabilize inorganic nanoparticles dispersed in a liquid medium, depending on the interactions between the particles and the polymer as well as on the interactions of these components with the surrounding medium. In the first case, the polymer molecules may adsorb at the particle surfaces and thus prevent agglomeration of the particles, which would cause precipitation. Here, the polymer-coated particles do not agglomerate because the interactions between the adsorbed layer and the surrounding medium are more favorable than the mutual interactions of the polymer layers surrounding the particles. The reverse effect may also occur; that is, nanoparticles can be destabilized by a polymer and precipitate spontaneously together with the polymer (coprecipitation). In stable mixtures of nanoparticles and polymer in a liquid medium, coprecipitation can also be induced by addition of a solvent that interacts only poorly with the polymer, thus urging the polymer to leave the solution, ideally together with the particles if the polymer–particle interactions are strong enough (it may also happen that only a part of nanoparticles precipitate together with the polymer).

When visually clear mixtures of *in situ*-prepared particles and polymer are present, uniform nanocomposite films are frequently obtained by casting followed by solvent evaporation, or by spin coating. The thicknesses of the films generated with these two methods are commonly in the range of 1 μm–1 mm. Polymers that are already known for good film formation properties may be preferred in these processes, such as poly(vinyl alcohol) for film preparation from aqueous mixtures or polystyrene when organic solvents are employed.

Finally, in some cases neat monomers can act as a solvent, and the nanocomposites are then formed first by *in situ* synthesis of the particles followed by *in situ* polymerization. Thus, there are many paths to embed *in situ*-synthesized particles into solid polymers, which as a whole offer chances for the preparation of completely different kinds of nanocomposites, together with other methods beyond the scope of this article, such as *in situ* synthesis of nanoparticles by decomposition of precursor compounds in solid polymer matrices [8] or by vapor phase techniques [9].

As a side remark, it should be considered that the nanoparticles in a particular sample are hardly of exactly one size. Therefore, the particle diameters addressed in the following always refer to average diameters, whereat the diameter distribution is often narrow (sometimes called "monodisperse" although a certain degree of size variation is evident upon careful evaluation).

2.2 METALS

Metal nanoparticles are conveniently synthesized *in situ* by reduction of metal centers, for example, in aqueous solutions by reduction of Ag^+ ions or Au^{III} as present in $[AuCl_4]^-$ ions. Reduction has been induced not only by common reducing agents but also by light or heat. On the other hand, organometallic precursors with metal atoms already present in the oxidation state of 0 were also applied, for example, $[Fe(CO)_5]$ or $[Co_2(CO)_8]$ in organic solvents. In such cases, nanoparticles are readily generated by release of the ligands, for instance, initiated by heat.

Attention has to be paid to the oxidation of metal nanoparticles. While experience has shown that nanoparticles of noble metals such as gold, silver, or platinum are not noteworthy affected by oxide formation, the synthesis of nanoparticles of metals that are prone to oxidation in the bulk state, such as iron, cobalt, or tin, should be performed under inert gas atmosphere. Otherwise, substantial amounts of metal oxides can readily arise in the products. When embedded in a polymer matrix, however, oxidation of nanoparticles is often markedly retarded, possibly by interaction of the polymers with the particle surfaces or by limitations of the diffusion of oxygen in the polymer matrix.

2.2.1 Gold

Among the metallic nanoparticles, those of gold have been the most widely studied. In fact, for centuries gold nanoparticles have been synthesized *in situ* in order to embed them in glass for their ruby color [10, 11] (which, however, turns to blue when the particle diameter increases; see succeeding text). Besides the colors of gold nanoparticles, the extraordinarily low refractive index of gold has also found attention. Remarkably, the related values are below 1.0 between wavelengths of 500 and 2000 nm [12], while the refractive indices of polymers lie typically in the range of 1.3–1.7. As a consequence, it is expected that the refractive index of nanocomposites decreases significantly with increasing volume fraction of embedded gold particles.

Noteworthy, polymer nanocomposites with *in situ*-synthesized gold particles were prepared already as early as in 1833 [13, 14]. A solution containing gold salts was reduced with Sn^{II} in the presence of gum Arabic, whereupon gold nanoparticles formed. The reaction mixture was clear, and ruby nanocomposites were subsequently produced by coprecipitation upon addition of ethanol.

At present, $Na[AuCl_4]$ and $H[AuCl_4]$ are frequently used as starting materials for the *in situ* synthesis of gold nanoparticles. These compounds are soluble in water and are then readily reduced by treatment with ordinary reducing agents, not only by Sn^{II} as mentioned earlier but also by, for example, sodium borohydride, hydrazine, or sodium citrate, to establish gold nanoparticles. Sodium citrate is also suited to control the particle diameter by variation of the ratio between sodium citrate and $[AuCl_4]^-$ [15].

Hence, a method to prepare nanocomposites with gold particles starts from treatment of $Na[AuCl_4]$ in hot aqueous solutions with different amounts of sodium citrate for 5 min. When gold dispersions thus generated have adopted room temperature, they are added to poly(vinyl alcohol) solutions [16, 17]. Subsequently, the solutions are poured into Petri dishes, and water is removed at a temperature of 80 °C and a pressure of 3 mbar. In this way, polymer films comprising gold nanoparticles of average diameters between 9.5 nm and 79 nm were obtained. Since poly(vinyl alcohol) is colorless, the color of the composites depended on the diameter of the gold particles: the films were red (9.5 nm diameter), purple (43 nm diameter), or blue (79 nm diameter). As it has been demonstrated by Mie and his Ph.D. student Steubing already 100 years ago, the colors of gold nanoparticles below c. 50 nm are caused mainly by absorption and not by scattering of light [18–20]. Accordingly, the colors of the red and purple poly(vinyl alcohol)–gold nanocomposites perceived in transmission and oblique observation did not differ considerably. In general, however, as gold particle diameters increase towards 50 nm, light scattering becomes more pronounced [18–20], resulting in different colors of the composites in transmission and oblique observation. Indeed, the nanocomposites

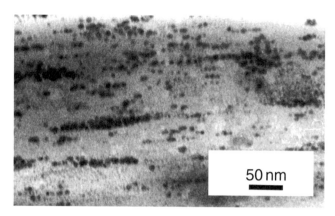

Figure 2.2. Transmission electron microscope picture of a nanocomposite consisting of dodecanethiol-coated silver nanoparticles and polyethylene after drawing at c. 120 °C (draw ratio 15) [21]. The picture contains linear assemblies of gold nanoparticles, whereat the assemblies are oriented parallel to the drawing direction.

comprising particles with 79 nm diameter appeared blue and brown when regarded in transmission and oblique position, respectively.

Poly(vinyl alcohol) belongs to the polymers that can be drawn at elevated temperatures, whereat polymer molecules orient in general along the drawing direction. The extent of drawing is given by the draw ratio, which designates the length ratio of the region subjected to drawing before and after drawing. The drawing ability of poly(vinyl alcohol) was retained in nanocomposites comprising *in situ*-prepared gold particles [16, 17]. Maybe somewhat unexpectedly, upon drawing at c. 120 °C, the embedded gold nanoparticles formed elongated assemblies aligned in the drawing direction at draw ratios of 5–6.

The driving force for the generation of the linear particle assemblies in these systems might be exclusion of particles from oriented crystalline polymer domains that establish upon drawing. Remarkably, the drawn materials showed a dichroism; that is, the color observed with linearly polarized light depended on the angle φ between the polarization plane of the light and the direction of the long axis of the particle assemblies ($\varphi = 0°$ denominates parallel and $\varphi = 90°$ perpendicular orientation). In the case of poly(vinyl alcohol)–gold nanocomposites, the color changed typically from blue at $\varphi = 0°$ to red at $\varphi = 90°$. Concomitantly, the UV–Vis absorption spectra recorded with linearly polarized light depended on φ. The absorption maximum arose at higher wavelengths for $\varphi = 0°$ compared to $\varphi = 90°$. Both orientation of particles upon drawing and dependence of UV–Vis absorption spectra on φ were also observed with dodecanethiol-modified silver particles in polyethylene after drawing (Figure 2.2 and Figure 2.3). Due to the color dependence on the position of the polarization direction of incident light, dichroic nanocomposites have been shown to act as color-providing component in bicolored liquid crystal displays [21, 22].

Other experiments for the creation of dichroic poly(vinyl alcohol)–gold nanocomposites started from water/ethylene glycol mixtures containing dissolved polymer and

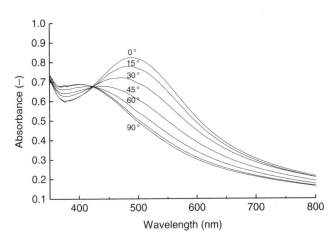

<u>Figure 2.3.</u> Visible light absorption spectra of a polyethylene–silver nanocomposite after drawing at c. 120 °C (draw ratio 15). The spectra were recorded with polarized light at various angles φ (0°–90° in 15° steps), whereat φ is the angle between the polarization plane of the light and the drawing direction [21].

H[AuCl$_4$] [23]. Films were then prepared by casting and subsequent evaporation of the solvents at ambient conditions. In these films, gold particles of average diameter around 5–10 nm were formed by irradiation with UV light, that is, without addition of a particular reducing agent. Remarkably, the mobility of the gold species in the solid polymer matrix was high enough to enable accumulation of gold atoms to particles. Particle formation was visually evident by a color change of the films to purple. At draw ratios of 4–5, a dichroism emerged, and the absorption maxima in UV–Vis spectra recorded with polarized light at φ=0° and φ=90° differed by up to 83 nm (typical values in the range of 20–40 nm). When poly(ethylene-*co*-vinyl alcohol) was applied as the matrix, water was not suited as a solvent as this copolymer is insoluble in water. Yet nanocomposites could be prepared from an organic solvent, dimethyl sulfoxide, in the same manner as described earlier with poly(vinyl alcohol) in aqueous solution. However, in the materials with the copolymer, the particle diameters gradually decreased from 15–20 nm at the surface to 2–3 nm in regions well below the surface. In spite of this, dichroism was observed after drawing, similar to the samples with poly(vinyl alcohol).

Interestingly, dichroic nanocomposites were also prepared in the end of the nineteenth century by *in situ* synthesis of gold nanoparticles in natural polymer matrices comprising uniaxially oriented hollow spaces [24, 25]. Thin specimens of spruce or animal fibrils were exposed to solutions of gold chlorides in water, ethanol, or aqueous formic acid (1%). Under the action of light, gold ions were reduced, thus producing linear particle assemblies due to the confinements of the hollow spaces in the fibrils. Other dichroic samples based on a natural polymer matrix were established by stretching of swollen gelatin comprising *in situ*-prepared gold nanoparticles and subsequent drying in the stretched state [26].

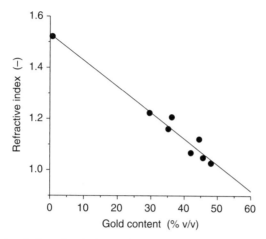

<u>Figure 2.4.</u> Refractive index of gelatin–gold nanocomposites as a function of the volume fraction of gold [27].

Gelatin was also employed as a matrix polymer for isotropic nanocomposite films with ultralow refractive index [27]. In order to prepare such materials, gelatin and H[AuCl$_4$] were treated first with acetone and then with water (acetone had to be employed first as otherwise part of the reaction mixture stuck to the walls of the reaction vessels). When the temperature was increased to 50–60 °C under stirring of the mixture, the solids dissolved. Thereafter, the solution was allowed to stand at room temperature for one minute, and hydrazine hydrate, an efficient reducing agent, was added. The gold ions were rapidly reduced and gold nanoparticles formed immediately. Films with gold contents ranging from 9.5% w/w (0.7% v/v) to 92.9% w/w (48% v/v) were prepared by spin coating of the reaction mixture on glass or silicon substrates (simple evaporation of the solvents resulted in films of minor quality). Average gold particle diameters amounted to about 5 nm in the films with the lowest and 50 nm in the films with the highest gold loadings. The films (thicknesses 200–400 nm) were initially mat and whitish, but after dipping into cold water, they adopted the blue or reddish-blue color of the gold nanoparticles. It was evident from scanning electron microscopy images that a crystalline compound at the surfaces of the films was removed by immersion into water, most likely a by-product of the reduction process that was not incorporated throughout the matrix but formed at the film surfaces. As mentioned earlier, gold exhibits a particularly low refractive index in a broad wavelength region, and accordingly the refractive index of the films decreased more or less linearly with the volume fraction of gold (Figure 2.4). Therefore, extraordinarily low refractive indices of 0.96 at 632.8 nm and 1.04 at 1300 nm were found in the nanocomposites for gold contents of 92.9% w/w (48% v/v).

The preceding examples illustrate that gold nanoparticles are readily accessible by reduction of [AuCl$_4$]$^-$ not only with common reducing agents but also by exposure to light. This allows to prepare nanocomposites by different methods, including generation of films by casting or spin coating. Since gold is not subjected to oxidation and the particles do not agglomerate once trapped in the solid matrix, the properties introduced by the gold nanoparticles into the nanocomposites (color, dichroism, low refractive index) persist for virtually unlimited periods.

2.2.2 Cobalt

Cobalt nanoparticles are rather sensitive to oxidation in the ambient, but oxidation is frequently retarded considerably when the particles are embedded in a polymer matrix. In fact, cobalt nanoparticles have attracted interest as a convenient representative of a superparamagnetic species [28]. As indicated earlier, superparamagnetism can emerge for particles with diameters below the critical size of magnetic domain wall formation, provided that most particles are well separated from each other.

A suited procedure for the synthesis of magnetic cobalt nanoparticles is based on thermolysis of $[Co_2(CO)_8]$ (dicobalt octacarbonyl) [29–32], which offers the advantage that only a volatile by-product (carbon monoxide) is released upon nanoparticle formation. When using $[Co_2(CO)_8]$ as the starting material for cobalt nanoparticle synthesis, attention has to be paid to its stability and handling [33]. Common commercial products consist of dark-violet crystals moistened with hexane. When exposed to the ambient, these crystals readily desiccate and adopt a whitish discoloration at their surfaces already after few minutes, probably due to oxidation products. When such materials are exposed to common organic solvents (e.g., toluene, p-xylene, decaline, hexane, or chlorinated benzenes), black, typically insoluble residues remain, which are not present when $[Co_2(CO)_8]$ from a freshly opened bottle is used. In fact, thermolysis of $[Co_2(CO)_8]$ appears to proceed at a slow rate already at room temperature. Hence, it is recommended to store $[Co_2(CO)_8]$ in a refrigerator prior to use and to employ $[Co_2(CO)_8]$ only from freshly opened bottles.

Since size and shape of cobalt particles are usually controlled by the interface between particles and adsorbing species present during particle formation [34, 35], agents believed to coordinate to the particle surfaces have been added to the reaction mixtures, in particular oleic acid, also in combination with trioctylphosphine oxide (TOPO) [30, 36]. Remarkably, the presence of TOPO during the preparation of cobalt nanoparticles yields the uncommon metastable ε-cobalt phase, which possesses a crystal symmetry like in β-manganese [29, 30, 32, 36].

Accordingly, cobalt nanoparticles were prepared by dissolution of $[Co_2(CO)_8]$ under an inert gas atmosphere at room temperature in toluene [33]. After addition of oleic acid and TOPO, the reaction mixture was rapidly heated to fast reflux and kept there for a few hours. During this time, $[Co_2(CO)_8]$ decomposed, yielding a homogeneous black liquid phase that contained cobalt nanoparticles of 4–5 nm diameter. Relatively high amounts of the stabilizing additives oleic acid and TOPO (mass ratio between oleic acid and cobalt of 5:1, molar ratio between oleic acid and TOPO of 5:1) were required to obtain homogeneous dispersions without agglomerated cobalt particles. If a magnet was approached to the reaction vessels, the reaction mixtures vigorously responded to the magnetic field, demonstrating the magnetic nature of the particles. Yet in the absence of TOPO, the reaction mixture did not respond to an external magnetic field, at least within a reaction period of 24 h.

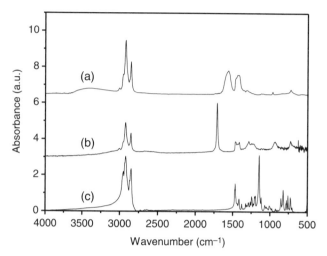

Figure 2.5. Infrared spectra of (**a**) cobalt nanoparticles synthesized in the presence of oleic acid and TOPO, (**b**) oleic acid, and (**c**) TOPO [33].

Subsequently, cobalt nanoparticles were precipitated from toluene dispersions by addition of ethanol [33]. In infrared spectra of the filtered and dried particles, two bands at $1557\,cm^{-1}$ and $1417\,cm^{-1}$ were detected (Figure 2.5), which are characteristic for the asymmetric and symmetric COO^- stretching vibration of oleate on cobalt [37], while the $C=O$ stretching vibration of the COOH group of oleic acid at $1710\,cm^{-1}$ and the characteristic peak of the $P=O$ group of TOPO at $1144\,cm^{-1}$ were absent, indicating that the excess of oleic acid and TOPO was efficiently removed by washing the particles with ethanol [33]. X-ray diffraction (XRD) measurements and electron diffraction studies revealed that the particles were present in the ε-cobalt crystal modification.

When the filtered and ethanol-washed particles were still wet, they could be re-dispersed to translucent liquids in hexane or toluene. However, re-dispersion in hexane or toluene was not complete when the samples were dried; obviously the particles agglomerate in the dry state in spite of adsorbed oleate. Hence, the freshly precipitated but not the dry particles have to be embedded in polymers. Thus, wet cobalt particles were employed for mixing with dissolved polychloroprene; this polymer turned out to be better suited for the preparation of uniform nanocomposites than other tested polymers [33]. The polychloroprene–cobalt solutions were spread on glass, and translucent films containing 9.9% w/w or 21.9% w/w cobalt resulted after solvent evaporation. Transmission electron microscope (TEM) images of these materials showed that the particles were randomly dispersed in the materials with 9.9% w/w cobalt but arranged partially into a pattern resembling rather a cubic than a hexagonal grid in the materials containing 21.9% w/w cobalt (Figure 2.6). The particle sizes did not depend on the polymer content; that is, the process of composite preparation did not lead to a significant change in particle diameter. Distinct peaks that could correspond to cobalt oxide were not detected with XRD, even after storing samples at ambient conditions for 3 months,

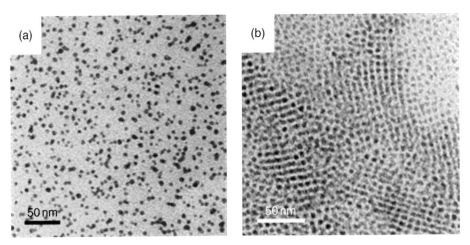

Figure 2.6. Cobalt particles (surface-modified with oleate) in polychloroprene. (a) 9.9% w/w cobalt and (b) 21.9% w/w cobalt [33].

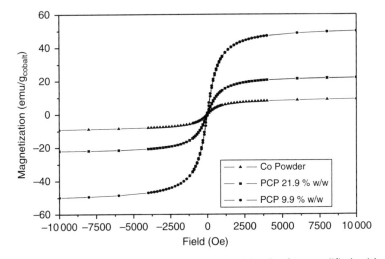

Figure 2.7. Magnetization curves of cobalt nanoparticles (surface-modified with oleate) as a powder and embedded in a polychloroprene (PCP) matrix (9.9% w/w and 21.9% w/w cobalt, respectively) [33].

although, of course, a certain degree of oxidation that does not manifest in the XRD patterns cannot be excluded. Excessive oxidation of cobalt might be retarded by coordination of oleate on the particle surface.

Importantly, magnetic measurements of the polychloroprene–cobalt nanocomposites with 9.9% w/w and 21.9% w/w cobalt showed a hysteresis loop that is typical for a superparamagnetic behavior (Figure 2.7) [33]. The magnetization curves at 300 K of freshly prepared composites showed the same coercivity of 15 Oe as the precipitated particles. The composite with 9.9% w/w particles possessed a saturation magnetization of 50.6 emu/g (the mass unit refers to the mass of cobalt) and that with 21.9% w/w particles a value of 35.3 emu/g, compared to 9.9 emu/g for the precipitated particles (containing 30% w/w cobalt and 70% w/w oleate). This indicates a reduction in magnetization per cobalt particle with increasing cobalt content in an organic matrix, in agreement with the behavior of cobalt particles in wax, which had been attributed to a modification of the exchange coupling between the adjacent particles upon dilution of the nanoparticles [29].

As a short summary, the experiments with cobalt nanoparticles show that also air-sensitive metal nanoparticles can be synthesized, that *in situ* modification of metal nanoparticle surfaces can be used to stabilize particle dispersions in liquid media, and that polymers can strongly retard oxidation of embedded metal nanoparticles. Also, the superparamagnetic behavior of cobalt particles becomes a property of the nanocomposites as a macroscopic entity, and it appears that dilution of superparamagnetic particles in an organic matrix might lead to an increase in magnetization per particle.

2.2.3 Platinum

Since platinum belongs to the noble metals, significant oxidation of platinum nanoparticles does not occur at ambient conditions, which renders this metal suited for model investigations (like gold). Accordingly, it will be shown in the following by means of platinum nanoparticles that unusual structures can be generated in a polymer matrix under certain conditions. As in the preceding case with cobalt, the platinum nanoparticles were synthesized by decomposition of a metal(0) complex and subsequently exposed *in situ* to an organic compound that binds to the particle surfaces, but in contrast to the experiments with cobalt, not only the particles but also the polymer was synthesized *in situ*.

With platinum, patterns were established by nanoparticle assemblies in a completely different manner than the just described linear assemblies of gold particles, and the structures of these assemblies also distinguish from nanoparticle patterns induced by microphase-separated block copolymers [38] or lattice formation of particles in highly concentrated dispersions [39, 40]. Self-organization of platinum nanoparticles was provoked by addition of sulfur-containing compounds with alkyl groups to *in situ*-prepared platinum particles, especially ammonium O,O'-dialkyldithiophosphates, $NH_4[S_2P(OR)_2]$ ($C_x$dtpNH$_4$, where x denotes the number of carbon atoms in each of the alkyl chains), or to some extent also with 1-octadecanethiol, $HS(C_{18}H_{37})$ [41]. These molecules are assumed to coordinate to platinum surfaces by sulfur atoms [42, 43], while the alkyl groups are expected to influence on the one hand the mutual interactions between the particles and on the other hand the interactions of the particles with the surrounding medium, which has impact to the dispersion behavior of the particles.

As a precursor for the platinum nanoparticles, tris(styrene)platinum(0), [Pt(styrene)$_3$], was prepared *in situ* by reduction of *cis*-[PtCl$_2$(styrene)$_2$] in styrene as a solvent (Figure 2.8) [41]. Then, a part of the styrene was removed at room temperature under reduced pressure (c. 0.01 mbar). During this process, [Pt(styrene)$_3$] decayed, as obviously a minimum excess of styrene must be present to stabilize the compound, leaving platinum atoms that agglomerated to particles of 1–2 nm diameter. The formation of the platinum particles was evident from a change of the solutions' colors from slightly yellow to brown. To these solutions, O,O'-dialkyldithiophosphates or 1-octadecanethiol was added in order to modify the particle surfaces *in situ*, followed by azobis(isobutyronitrile) (AIBN), which is a common starting agent for the radical polymerization of styrene at 60 °C. Consequently, polystyrene was synthesized under the action of AIBN, and the reaction mixture slowly solidified. The number-average molar masses (M_n) of the final polystyrene amounted typically to 30000–40000 with M_w/M_n around 5–7, where M_w denotes the weight average molar mass. The size of the platinum particles and the molar masses of the polystyrene did not depend significantly on the presence of the sulfur-containing additives, as expected as the additives were applied only after particle formation.

The nanocomposites thus produced contained superstructures constituted by platinum nanoparticles, as evident from TEM. Remarkably the superstructures consisted of hundreds or thousands of platinum nanoparticles. The shape of the structures depended on the type of additive (i.e., $C_x$dtpNH$_4$ or 1-octadecanethiol), the ratio between platinum atoms and additive molecules (1:1–4:1), and the platinum content in the system (1–8% w/w). In the absence of additive, no patterns arose. Obviously, the formation of superstructures was governed by the alkyl chains of the additive molecules. As an example, onion-like superstructures composed of several shells or of single shells established by platinum particles emerged in TEM images of nanocomposites prepared in the presence of C_{18}dtpNH$_4$ or C_{12}dtpNH$_4$ at platinum/$C_x$dtpNH$_4$ ratios of 2:1 and 3:1 (Figure 2.8; note that such spheric structures appear as circles in the projections represented by the TEM pictures). These structures were characterized by "rings" of typical diameters of 50–300 nm and borderline thicknesses of 5–15 nm. Features composed of connected shells were also visible.

At first glance, the TEM images of the onion-like structures formed by the platinum particles resemble the superstructures of bilayer membranes (vesicles) formed by self-organization of organic molecules in solution [44–48]. Such membranes can also conjoin to assemblies such as networks [45], as also observed in the form of connected particle shells in nanocomposites of polystyrene and platinum in the presence of C_{18}dtpNH$_4$ or C_{12}dtpNH$_4$. Notably, the molecules forming bilayer membranes in solution are commonly composed of a polar head group that is connected to two alkyl groups comprising typically 12–18 carbon atoms. Hence, these membrane-forming molecules are of a similar type as the long chain O,O'-dialkyldithiophosphates. Although complexation of the head group by metal atoms has been reported to stabilize bilayer aggregates [44], there are striking differences between bilayer membranes and the superstructures formed by platinum atoms in polystyrene: typical bilayer membranes are formed in aqueous solutions as a consequence of a balance between hydrophilic and hydrophobic interactions, and the thickness of the membrane walls is molecularly controlled. In the polystyrene–platinum systems, the medium is apolar and the walls of the structures vary somewhat in thickness, which extends over a few particle diameters. Essentially, crystallization of surface-bound alkyl groups between adjacent particles might act as the driving force for the formation of platinum superstructures. The

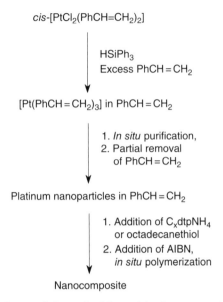

cis-[PtCl$_2$(PhCH=CH$_2$)$_2$]

HSiPh$_3$
Excess PhCH=CH$_2$

[Pt(PhCH=CH$_2$)$_3$] in PhCH=CH$_2$

1. *In situ* purification,
2. Partial removal
   of PhCH=CH$_2$

Platinum nanoparticles in PhCH=CH$_2$

1. Addition of C$_x$dtpNH$_4$
   or octadecanethiol
2. Addition of AIBN,
   in situ polymerization

Nanocomposite

Figure 2.8. Reaction scheme of the embedding of *in situ*-prepared platinum nanoparticles with *in situ*-modified surfaces into *in situ*-synthesized polystyrene [41]. C$_x$dtpNH$_4$ designates ammonium O,O'-dialkyldithiophosphates.

crystallization ability probably depends on many parameters, in particular on the surface coverage of the surface-bound molecules, the concentration of the nanoparticles in the matrix, and the length of the alkyl groups. In spite of the differences between membranes and platinum superstructures, the existence of concentric shells in the latter system, such as shown in Figure 2.9, could be explained analogously to related phenomena in bilayer membranes [48] based on shape variations of a given shell with time (Figure 2.10).

Upon addition of octadecanethiol, C$_6$dtpNH$_4$ or C$_2$dtpNH$_4$ to the platinum particles, other features were observed in the final polystyrene matrix than with C$_{18}$dtpNH$_4$ or C$_{12}$dtpNH$_4$ (see detailed descriptions in the literature [41]). This underlines that the chemical nature of the head group and the length of the alkyl groups of the additive molecules, which are probably bound to the particle surfaces, are of importance for pattern formation. Not only the extent of the interaction of the sulfur-based head group with the platinum surfaces is of importance for the number of surface-bound alkyl groups per surface area but also the space requirement of the head group and the alkyl groups. Essentially, the length of the alkyl groups is expected to determine the balance between crystallization of alkyl groups and interaction of the organic particle shell with the matrix.

As a consequence of the aforementioned findings, it has become evident that appropriate modification of surfaces of *in situ*-prepared nanoparticles with organic molecules may be used not only to stabilize particles in dispersions but also to generate superstructures constituted by particles in the final nanocomposites.

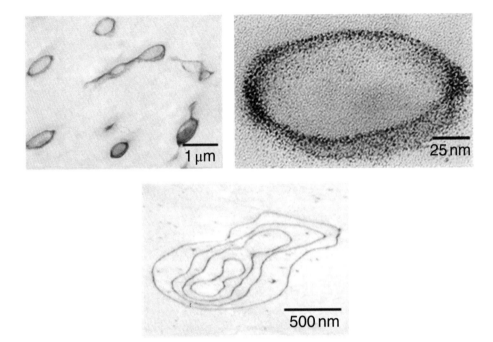

Figure 2.9. Superstructures formed by platinum nanoparticles in polystyrene, established by *in situ* polymerization of styrene in the presence of ammonium *O,O'*-dioctadecyldithiophosphate [41].

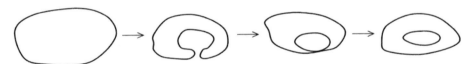

Figure 2.10. Sketch of a possible formation of concentric shells of platinum nanoparticles in a polymer matrix upon *in situ* polymerization of styrene in the presence of ammonium *O,O'*-dioctadecyldithiophosphate [41].

2.3 METAL OXIDES

Nanoparticles of metal oxides are conveniently prepared via hydrolysis of suited precursors, for instance, $TiCl_4$, $Ti(O^iPr)_4$ (^iPr = isopropyl), or $Zn(C_2H_5)_2$ to produce TiO_2 or ZnO, respectively. When such precursors are applied, volatile by-products arise, namely, HCl (which may, however, induce degradation reactions of susceptible polymers), isopropanol, or ethane, respectively. In fact, it may be of advantage to select precursors that yield volatile by-products as these can evaporate during drying of the final nanocomposites. This is of advantage, for example, if films, prepared by casting or spin coating, are desired without incorporation of by-products.

2.3.1 Titanium Dioxide

Titanium dioxide nanoparticles find special attention in the area of nanocomposites as they possess a rare and attractive combination of optical properties: a high refractive index (2.5–2.7, depending on the crystal modification), transparency in the Vis wavelength region, and pronounced absorption of UV radiation close to the Vis wavelength region. Notably, transparent materials with high refractive indices are of interest in areas such lenses, reflectors, optical filters, and optical waveguides, while absorption of UV radiation is essential in UV-protective coatings.

There are different polymorphic crystal structures of TiO_2, the most important of which are rutile and anatase. These crystal modifications can differ not only in their physical properties but also in their chemical behavior; for instance, it is generally accepted that the photoactivity of anatase is typically more pronounced than that of rutile [49–51]. Although rutile is the thermodynamically stable crystal modification of TiO_2, anatase is typically the modification that emerges in nanoparticles synthesized *in situ* in liquid media, for kinetic reasons. However, rutile nanoparticles may arise when $TiCl_4$ is hydrolyzed at low temperatures in strongly acidic environment. The hydrolysis of $TiCl_4$ follows the equation

$$TiCl_4 + 2\,H_2O \rightarrow TiO_2 + 4\,HCl$$

This reaction equation appears simple, but nanoparticle formation proceeds via a complex sequence of reaction steps that have not been identified in full detail [52, 53]. When TiO_2 is prepared by addition of $TiCl_4$ to water, particle size and finally resulting crystal structure can depend on numerous parameters, such as the initial concentration of $TiCl_4$, addition rate of $TiCl_4$, reaction temperature, and pH value.

As an example, rutile nanoparticles were synthesized *in situ* by addition of freshly distilled $TiCl_4$ under argon atmosphere to an ice-cooled aqueous HCl solution (note that HCl was added, although HCl was also released upon TiO_2 formation) [54]. Thereafter, the solution was kept at 60 °C, whereupon initially formed titanium oxide hydrates slowly developed to TiO_2 nanoparticles. Solutions thus obtained were mixed with aqueous solutions of poly(vinyl alcohol), poly(vinyl acetate) 88% mol/mol hydrolyzed, poly(vinylpyrrolidone), or poly(4-vinylpyridine). Some of the dispersions were neutralized with sodium carbonate, and the sodium chloride generated in this process was subsequently removed by dialysis, which took, however, many days. Finally, nanocomposite films were prepared by casting the transparent neutralized or non-neutralized TiO_2 dispersions, followed by evaporation of the water at a temperature of 30 °C and a pressure of c. 100 mbar. The particle diameters were in the range of 2.5 nm, and rutile was the only crystalline compound that could be detected by electron diffraction. Yet in general, samples with *in situ*-prepared TiO_2 nanoparticles may also contain considerable amounts of titanium oxide hydrate particles, the amorphous precursor to TiO_2, which can hardly be detected with electron diffraction or XRD of poly(vinyl alcohol)–TiO_2 films (thicknesses 40–80 μm) with TiO_2 contents of 2–25% w/w showing particles well dispersed in the matrix. The samples appeared colorless to the eye and showed in UV–Vis spectra little absorbance in the Vis wavelength range but strong absorption in the UV

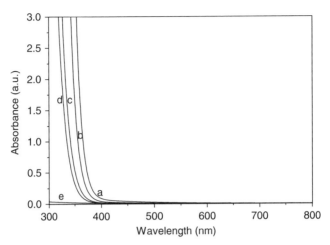

Figure 2.11. UV–Vis absorption spectra of poly(vinyl alcohol)–TiO_2 nanocomposites at different mass contents (% w/w) **(a)** 23.0, **(b)** 11.0, **(c)** 6.9, and **(d)** 4.6 and **(e)** pristine poly(vinyl alcohol) [54].

region (Figure 2.11) [54]. For instance, UV light was efficiently absorbed below c. 360 nm in materials with 23% w/w TiO_2, which therefore act as visually transparent broadband UV filters. After a few days in the ambient, however, samples that had not been subjected to the aforementioned neutralization procedure adopted a yellowish discoloration, which also appeared in reference samples of poly(vinyl alcohol) itself, processed in acidic environment in the same manner as the nanocomposites. It seems, therefore, that residual HCl from the particle synthesis procedure leads to degradation of poly(vinyl alcohol).

Concerning the employed polymers other than poly(vinyl alcohol), composite films with 88% hydrolyzed poly(vinyl acetate) showed transparency up to TiO_2 fractions in the range described earlier with poly(vinyl alcohol). However, films with poly(vinylpyrrolidone) or poly(4-vinylpyridine) as the matrix virtually exhibited transparency to the eye only up to TiO_2 fractions of 4–6% w/w. Due to the lower TiO_2 content, films of 100 μm thickness based on the latter two polymers absorbed UV light efficiently only below c. 300–320 nm.

As mentioned earlier, besides UV absorption capacity, the high refractive index of TiO_2 is another subject of interest. Accordingly, the incorporation of TiO_2 into poly(vinyl alcohol) resulted in an increase of the refractive index, which augmented linearly with the volume fraction of TiO_2 (Figure 2.12), reaching a value of 1.609 at 10.5% v/v. A linear increase of the refractive index on the volume fraction of particles was also observed in other systems (see succeeding text). Extrapolation of the refractive indices of the poly(vinyl alcohol)–TiO_2 nanocomposites to 100% TiO_2 resulted in a refractive index of 2.30, which is below the aforementioned values for crystalline TiO_2. This deviation could be due to a nonlinear dependence of the refractive index of the composite on the volume content of particles at high particle loadings [55–58], a decrease of the

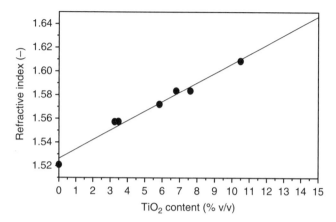

Figure 2.12. Refractive index dependence of poly(vinyl alcohol)–TiO$_2$ nanocomposites on the TiO$_2$ content [54].

refractive index below a certain particle diameter (see succeeding text), or the presence of titanium oxide hydrates in the nanoparticle fraction.

As another interesting materials property, it has been reported that TiO$_2$ suspended in liquids shows a photochromic effect in the presence of certain organic compounds such as alcohols, due to partial reduction of Ti^{IV} to Ti^{III}, which leads to a color change to blue [59, 60]. However, in the dispersions in liquid media, the blue color rapidly faded away under ambient conditions when the light source was removed, as a consequence of reoxidation of Ti^{III} to Ti^{IV}. Photochromic behavior was also observed when TiO$_2$ was embedded in poly(vinyl alcohol), but in this case the blue color persisted for long periods [61, 62]. Hence, blue films (c. 100 nm thickness) with *in situ*-prepared TiO$_2$ particles of 3 nm diameter kept their colors at least for weeks at ambient conditions [62]. Further, when initially colorless films were irradiated through a mask, color patterns with a resolution of 10 μm or less could be generated. Remarkably, by wetting with water, blue films became colorless again. It appears that oxygen can penetrate into the water-swollen matrix, thus getting into contact with the particles and oxidizing Ti^{III} to Ti^{IV}; that is, the photochromism is reversible at special conditions. Yet it has to be considered that recent investigations imply that the photochromism in systems with TiO$_2$ could be due rather to titanium oxide hydrates present in the particles than to TiO$_2$ itself [63]; note that titanium oxide hydrates are common reaction intermediates in the *in situ* formation of TiO$_2$ and it may readily happen that the conversion of titanium oxide hydrate particles to TiO$_2$ is not complete.

The aforementioned findings show that indeed TiO$_2$ particles embedded in polymers can combine visual transparency with other optical properties such as pronounced UV absorption and high refractive index. However, it should be considered that upon *in situ* formation of TiO$_2$, residual titanium oxide hydrates can be present that also contribute to the properties of the nanocomposites or could even cause effects frequently ascribed to TiO$_2$, as in the case of photochromism.

2.3.2 Zinc Oxide

Similar to TiO_2 (see preceding text), ZnO nanoparticles embedded in polymer films strongly absorb UV radiation from c. 360–380 nm down to the common detection limit of UV–Vis spectrometers (about 200 nm), while they are highly transparent in the visual wavelength region [64–67]. Such particles (average diameters on the order of 5–10 nm) were synthesized *in situ* by reaction of zinc acetate dihydrate with sodium hydroxide in ethanol [67]. After addition of the monomer butanediol monoacrylate, removal of ethanol, ageing of the reaction mixture, and separation from precipitates, the resulting dispersion was supplemented with the cross-linking agent trimethylolpropane triacrylate. Subsequently, nanocomposite films were generated by pouring the reaction mixture between two glass plates separated by 1.5 mm followed by radical polymerization of the acrylates, initiated by UV light and a photoinitiator. Indeed, samples up to c. 8% w/w ZnO were transparent in the Vis wavelength region and efficiently absorbed UV light below wavelengths of c. 350–360 nm.

Commonly, the ZnO nanoparticles are distributed homogeneously throughout the matrix. However, ZnO nanoparticles were also generated *in situ* exclusively in a surface region of the matrix by means of diffusion processes [68]. For this purpose, diethyl zinc is a well-suited starting compound as it readily hydrolyzes [69], according to the equation

$$\left(C_2H_5\right)_2 Zn + H_2O \rightarrow ZnO + 2\,C_2H_6$$

Only ethane is left as a reaction by-product, which is volatile and can therefore readily escape from the matrix. Thus, hexane solutions of diethyl zinc were applied for the treatment of sheets of a statistical copolymer of ethylene and vinyl acetate (EVA), which is accessible to penetration of hexane [68]. For the preparation of ZnO surface layers, the diffusion rate of diethyl zinc in hexane was found to be most appropriate at temperatures between −15 and −30 °C. At −50 °C the diffusion rate was too low to incorporate a significant quantity of diethyl zinc in the polymer within an immersion time of 2 h, while around 0 °C the diffusion was so fast that it was difficult to control the penetration depth of diethyl zinc. Yet, for instance, after immersion of EVA sheets into a solution of diethyl zinc (0.167 M) at −15 °C for 25–40 min and subsequent exposure to air, the incorporated diethyl zinc was converted by hydrolysis with the ambient humidity to ZnO, and the absorbed hexane evaporated. Under such conditions, the ZnO particles were confined in a surface region of 1–2 μm thickness (on each of the two surfaces of the films), as evident from electron micrographs of cross sections of ZnO-containing EVA sheets. Within this layer, a content of 5–10% v/v ZnO was estimated after analysis of the zinc content by atomic emission spectroscopy. The particles that finally formed were of c. 15 nm in diameter and consisted of zincite.

Figure 2.13 shows UV–Vis spectra of an EVA sheet with a surface layer of ZnO nanoparticles and, for comparison, of an untreated EVA sheet as received. The spectra disclose that some light in the UV region is absorbed by EVA alone, either by EVA itself or by impurities in EVA, but the pronounced UV light absorption of the sample with

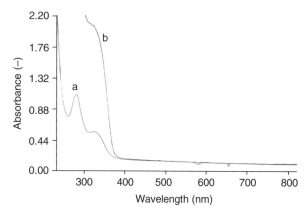

Figure 2.13. UV–Vis absorption spectra of (a) neat poly(ethylene-co-vinyl acetate) and (b) EVA with a layer of ZnO in the surface region [68].

ZnO layer demonstrates clearly that the quantity of ZnO is sufficient to absorb UV light efficiently below 380 nm, while in the range of Vis wavelengths, the ZnO nanoparticles did not absorb light significantly. Accordingly, these samples appeared completely transparent to the eye.

The aforementioned experiments show that diffusion of particle precursors in swollen polymers can indeed be utilized to create a layer comprising *in situ*-generated particles only in the surface region of bulk polymers. This layer can be sufficient to introduce materials properties that are not characteristic for the polymer itself (UV absorption in the example of ZnO). Importantly, however, in such materials bulk properties of the polymer are expected to remain unaffected also at highest particle concentrations in the surface region. For instance, increased brittleness of the bulk material, which frequently occurs when a considerable fraction of particles are homogeneously distributed in the entire volume of a matrix, may thus be avoided.

2.4 METAL SULFIDES

Metal sulfides are conveniently synthesized in aqueous solutions by combination of metal aqua ions and a sulfide source, such as Na_2S, NaHS, or H_2S. The latter offers the advantage of volatile by-product formation if metal acetates are used as starting materials since acetic acid is then formed as a reaction by-product upon metal sulfide generation. However, H_2S is more difficult to handle than Na_2S or NaHS because H_2S itself is a gas that is connected with certain additional efforts in experimental realizations; in addition, H_2S solutions typically release some gaseous H_2S (which can easily be identified by its characteristic odor). On the other hand, the application of Na_2S leads to an increase in the pH value, and therefore it has to be assured that precipitation of metal hydroxides does not occur.

2.4.1 Lead Sulfide

As mentioned earlier, the refractive index of polymers is essentially confined between 1.3 and 1.7. In marked contrast, PbS exhibits refractive indices as high as 4.0–4.6 in the wavelength range of c. 420 nm–11 μm [12]. Therefore, although PbS absorbs Vis light significantly, embedding of PbS nanoparticles in polymers is anticipated to be particularly suited for fundamental investigations with regard to the refractive index of nanocomposites, all the more the synthesis of PbS nanoparticles proceeds straightforward.

Lead sulfide nanoparticles were synthesized with lead acetate and hydrogen sulfide as the starting materials, according to the reaction

$$Pb(OAc)_2 + H_2S \rightarrow PbS + HOAc \, (Ac = acetyl)$$

Hence, acetic acid is the only by-product that goes along with PbS particle formation. Since acetic acid is liquid at room temperature, it is basically volatile and can thus be removed from the reaction mixture by evaporation or by filtration of precipitated composites. X-ray photoelectron spectra indicated that oxidation of PbS to $PbSO_3$ or $PbSO_4$, which can proceed within hours in aqueous dispersions, was not significant anymore when the particles were finally embedded in the polymer matrix. Also, the size distribution of embedded PbS particles did not change within some weeks.

Nanocomposites with high content of PbS (88% w/w, c. 50% v/v) were obtained by addition of an aqueous H_2S solution to a deoxygenated solution containing lead acetate and poly(ethylene oxide) (deoxygenation of water was performed by vacuum/N_2 treatment in order to avoid formation of sulfite or sulfate) [70]. PbS formed instantly and precipitated rapidly together with the polymer. Thus, poly(ethylene oxide) destabilized the particle dispersion, and coprecipitation was induced by formation of the PbS particles; that is, a precipitating solvent was not required. This implies an interaction between the PbS particles and the poly(ethylene oxide), maybe by coordination of oxygen atoms of the polymer with lead ions at the surface of the PbS nanocrystals, but also interactions between oxygen atoms of the polymer and sulfur atoms of PbS cannot be excluded [71]. The freshly precipitated products were slimy. In order to dry the materials, water was first removed by squeezing and kneading the composites between filter paper and afterwards by pressing of the solids between dialysis membranes surrounded by blotting paper. Finally, residual water was withdrawn with the help of a hydraulic press, whereupon the water content was finally reduced to c. 0.7% w/w. The average PbS particle diameter in these materials amounted to c. 15 nm. Remarkably, refractive index values of the composites between 2.9 and 3.4 (the differences reflecting variations between the samples) were found by ellipsometric measurements at wavelengths of 632.8 nm and 1295 nm, and typical values of 3.1–3.2 resulted from reflectivity measurements between 1000 and 2500 nm. These values are extremely high and have not been achieved with pure polymers so far, according to our knowledge.

In order to prepare films with a wide concentration range of embedded PbS nanoparticles, an aqueous solution comprising gelatin and lead acetate was treated with an aqueous H_2S solution [72]. PbS nanoparticles of dimensions below 10 nm formed immediately, but in contrast to poly(ethylene oxide), gelatin stabilized the particle

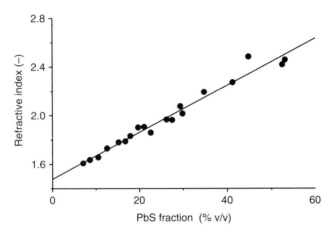

Figure 2.14. Refractive index of gelatin–PbS nanocomposites as a function of the volume fraction of PbS [72].

dispersion. Moreover, under the action of the gelatin, which had been dissolved at elevated temperature (50–60 °C), the viscosity of the reaction mixture increased within an hour at room temperature. As a result, the consistence of the reaction mixture became suited for the preparation of films by spin coating, which yielded more uniform samples at higher PbS contents than obtained by casting followed by solvent evaporation at ambient conditions, elevated temperatures, or reduced pressure. The cast films were uniform only up to c. 65% w/w PbS. Above this particle fraction, local phase separation was observed, and the film thicknesses were not uniform but decreased with increasing distance from the centers of the films. On the other hand, by spin coating films of thicknesses between c. 40 nm and 1 µm were prepared by variation of the rotation rate of the substrates. Homogeneous films up to 86.4% w/w PbS could be obtained; however, films with less than c. 70% w/w PbS were of significantly better quality. After annealing at 200 °C and 133 mbar for 2 h, the films shrank by 5–8%. No indications for the formation of heterogeneities by the annealing process were evident. The annealing led to an increase in the refractive indices of 1–8%, accompanied with a decrease in the deviations of the measured values. At a PbS content of 86.4% w/w (c. 50% v/v), an extraordinarily high refractive index of 2.47 was found at 632.8 nm in annealed samples. Also, the refractive indices increased within the experimental precision linearly with the volume fraction of PbS (Figure 2.14). Besides the aforementioned systems of poly(vinyl alcohol) and TiO_2, a linear increase of the refractive index with increasing volume fraction of particles was also found, for instance, in the case of a polyimide with embedded TiO_2 nanoparticles (highest refractive index 1.82) [73], a copolymer of vinyl monomers and ZnS (highest refractive indices 1.60–1.65) [74], or gelatin with silicon (highest refractive index 3.3) [58]. Indeed, a linear dependence of the refractive index on the volume fraction of embedded nanoparticles might be expected for many polymer composites and blends based on theoretical considerations, but nonlinear relations are basically also possible [55, 57, 58, 75, 76]. In fact, some deviation from a linear relationship might also be consistent with the experimental data of the aforementioned systems because of the limited precision of the refractive index measurements as well as of the calculated volume fractions.

It is known that suited chemical agents can be employed for the manipulation of the sizes of inorganic particles as they can influence basic steps in nanoparticle formation. Generally, the growth of nanoparticles in solution is assumed to proceed in two main steps: nucleation and particle growth. In such models, the particle sizes are established essentially on the basis of the rate ratios of these two main steps. If the nucleation rate can be accelerated relative to the growth rate, more nuclei form and accordingly less ions are available per particle that then results in a decrease in particle size, and vice versa for relative acceleration of the growth rate. In fact, the diameters of the PbS particles could be varied by addition of either sodium dodecyl sulfate or acetic acid to an aqueous poly(ethylene oxide)–lead acetate solution [77]. Upon addition of a H_2S solution, again nanocomposites formed by coprecipitation. Depending on the quantities, sodium dodecyl sulfate induced a decrease in PbS particle size down to 4 nm and acetic acid an increase up to 80 nm, while particle diameters of 29 nm resulted in the absence of these additives under the applied experimental conditions. Notably, Pb^{2+} ions basically interact with dodecyl sulfate as evident by a clouding after addition of dodecyl sulfate to the Pb^{2+}-containing solutions. Hence, the reactions between Pb^{2+} and H_2S that finally lead to PbS interfere with equilibria involving dodecyl sulfate, such as adsorption equilibria of dodecyl sulfate at growing particle surfaces, which could indeed passivate the PbS surfaces and therefore decrease the particle growth rate. Acetic acid, on the other hand, appears to increase the growth rate relative to the nucleation rate, maybe as a result of acid–base reactions, which are essential in PbS formation.

When inorganic particles are applied for refractive index modifications of nanocomposites, it has to be considered that the refractive index of the particles may change in an unfavorable way when the particles' diameters fall below a certain limit. This was demonstrated on the example of poly(ethylene oxide)–PbS composites. The refractive indices of composites with PbS diameters above c. 25 nm amounted to 3.5–3.8 at a wavelength of 632.8 nm and to 3.3–3.7 at 1295 nm [77]. Yet in the case of materials containing PbS particles of lower diameter, the refractive indices decreased, for example, to 1.7–1.8 at both wavelengths at particle diameters of 4 nm (Figure 2.15). The refractive indices of the PbS particles themselves were estimated from the values of the composites by linear extrapolation to 100% v/v PbS (cf. dependency of refractive index of nanocomposites and volume fraction of embedded particles discussed earlier). Two different regimes were observed here: above c. 25 nm particle diameter, the refractive indices extrapolated for PbS amounted to 4.0–4.3 at a wavelength of 632.8 nm and 3.8–4.1 at 1295 nm. These values were independent on the particle size within the experimental precision and lay in the region of the values of bulk PbS (4.19 at 619.9 nm and 4.275 at 1300 nm [12]). However, below c. 25 nm particle diameter, the extrapolated refractive indices of the PbS particles decreased, as expected from theoretical estimations, and refractive indices on the order of 2.0–2.3 were found for particles of 4 nm diameter at both wavelengths. Hence, when PbS nanoparticles are employed for the enhancement of the refractive index in composites, particle sizes below c. 25 nm should be avoided since the enhancing effect of the particles becomes less effective.

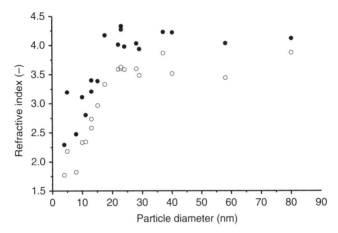

Figure 2.15. Refractive index of nanocomposites of poly(ethylene oxide) and PbS particles of different diameters (open circles) and extrapolated refractive indices of PbS (filled circles), at a wavelength of 632.8 nm [77].

When all the aforementioned experiments are considered, it is obvious that *in situ*-synthesized PbS nanoparticle dispersions can be stabilized or destabilized by polymers that are present during particle synthesis. In the former case, the materials precipitate spontaneously and can be isolated as homogeneous materials with high volume fractions of PbS by filtration or decantation of the supernatant liquids. On the other hand, stabilized dispersions can be used for the preparation of films. Importantly, the size of PbS particles can be varied by additives, and concomitantly their refractive index decreases below a certain size limit. This implies that basically particle size effects should be considered in any case of refractive index modification of polymers by embedding inorganic nanoparticles of extreme refractive index. The particles sizes should not be too low if the refractive index of nanocomposites is attempted to be modified drastically by inorganic nanoparticles.

2.4.2 Iron Sulfides

Besides PbS (see preceding text), also iron sulfides can exhibit very high refractive indices, as reflected by the value of pyrite (cubic FeS_2), which amounts to at least 3.5 (at 632.8 and 1295 nm) [78]. However, the chemistry of iron sulfides is much more complex than that of PbS. Depending on the experimental conditions, the stoichiometry of *in situ*-formed iron sulfides may vary, and apart from pyrite species such as "amorphous FeS" (the nature of which is controverse when comparing different reports [79, 80]), greigite (Fe_3S_4), mackinawite (FeS_{1-x}), pyrrhotite ($Fe_{1-x}S$), marcasite (orthorhombic FeS_2), and troilite (FeS) can arise [79, 81, 82]. Mackinawite was reported to be the most frequent product of the precipitation of Fe^{2+} ions by H_2S, NaHS, or Na_2S below 100 °C in the absence of oxidants [81].

As in the case of PbS, poly(ethylene oxide) was selected as matrix polymer for the preparation of nanocomposites by coprecipitation. In an aqueous poly(ethylene oxide) solution, an iron(II) salt was dissolved and a sulfur source was added, whereupon iron sulfides formed *in situ* and nanocomposites precipitated [78]. As an example, iron sulfides were synthesized *in situ* from $FeSO_4$ and H_2S at 75 °C, and the polymer precipitated after 20 h essentially quantitatively together with the iron sulfides. This is reflected by the linear dependence of the polymer fraction in the composites on the polymer concentration in solution at a constant quantity of dissolved $FeSO_4$, yielding materials with a polymer content between 1.5 and 15% w/w (corresponding to 4–5 and 30–40% v/v, respectively). The particles were so small and their content in the composites was so high that the particle diameters could not be determined reliably from TEM investigations. The nanocomposites exhibited refractive indices around 2.0. A distinct dependence of the refractive index on the composition was not observed, which might be due to different compositions of the inorganic components in the composites (various iron sulfides were identified with XRD) or to an increased pore volume in the pressed samples with higher loadings of inorganic particles.

Materials with higher refractive indices, namely, 2.5–2.8 (at 1295 nm), were prepared with $(NH_4)_2Fe(SO_4)_2 \cdot 6H_2O$ (Mohr's salt) and H_2S or NaHS [78]. The highest refractive indices were obtained with reaction times of several days and elevated reaction temperatures. Mohr's salt seemed to be more favorable for the preparation of iron sulfide nanoparticles providing a high refractive index to the composites than $FeCl_2$ or $FeSO_4$. This might be due to pH changes by ammonium ions in Mohr's salt, as the formation of the different iron sulfides can depend on the pH value. The samples with the highest refractive indices contained mackinawite and greigite and occasionally also pyrite, but it cannot be excluded that large fractions of "amorphous iron sulfides" were also present. It has to be considered that the nanocomposites most likely contained pores that may take up about 10% of the total volume. Pores could probably be avoided by increasing the polymer fraction; however, an excess of polymer will decrease the refractive index of the composite. A polymer content of roughly 10% v/v appeared to be a suited compromise for the endeavor to decrease the polymer and the pore fraction.

Thus, as a short recapitulation, iron sulfide nanoparticles vary considerably in stoichiometry when the reaction conditions change or when different Fe^{II} sources are employed, although it may appear at first glance that simply FeS should form when a soluble Fe^{II} compound is combined with H_2S or HS^-. Nonetheless, the various iron sulfides appeared to precipitate quantitatively together with poly(ethylene oxide), which allows to adjust the polymer fraction in composites in a simple way and to receive materials with high refractive indices.

2.5 CONCLUSIONS

Inorganic nanoparticles of different chemical compositions can be synthesized *in situ* in liquid media with readily available starting compounds in straightforward procedures. Often, nanoparticle formation follows an obvious overall reaction scheme, as, for example, in the generation of gold, ZnO, or PbS. However, particle formation is basically a complex process, and therefore special or even unforeseen product compositions may arise, as with cobalt (ε-cobalt), TiO_2 (rutile or anatase), or iron sulfides (various species). In principle, *in situ* synthesis of nanoparticles can be performed not only in water but also in organic solvents, which allows an adaption to the solubility of the polymer selected for the embedding of the particles. While nanoparticle synthesis in water is more established in several cases, for example, for gold, TiO_2 and PbS, relatively few of the common polymers dissolve in water. In order to enable mixing of water-insoluble polymers with particles prepared *in situ* in aqueous solutions, *in situ* modification of particle surfaces may change their solubility behavior and subsequently enable mixing of particles and polymers in organic solvents.

Mixing of *in situ*-synthesized particles and polymers in a liquid medium may be accompanied by simultaneous precipitation of particles and polymer (coprecipitation). If this way is desired for particle embedding in a polymer matrix, a polymer can be employed that destabilizes the particles in the liquid environment, or coprecipitation can be induced by treatment of the particle–polymer–solvent mixture with a precipitating agent (i.e., a non-solvent for the polymer). If, on the other hand, a polymer is used that stabilizes the particle dispersion in a liquid medium, the opportunity of nanocomposite film preparation is offered by casting or spin coating. In order to prevent precipitation of the particles, *in situ* modification of their surfaces with appropriate organic molecules of low molecular mass can also be helpful and moreover improve particle dispersibility in the final polymer matrix.

Further processing of nanocomposites by drawing can lead to anisotropic particle structures in the matrix, such as elongated particle assemblies that are oriented in the drawing direction. If the nanoparticles are colored, the respective composites show a dichroism and can act, therefore, for example, as color-providing component in liquid crystal displays. Exceptional structures formed by nanoparticles embedded in polymers may also arise under the action of surface-modifying agents, as shown with *in situ* surface-modified platinum particles in polystyrene.

Concerning materials properties, in general particles with dimensions well below the wavelength of incident light scatter light only very moderately if they are not considerably agglomerated. Thus, nanocomposites are readily transparent or translucent compared to their counterparts with larger particles (unless the refractive indices of particles and polymer are equal, which always leads to transparent materials). Furthermore, nanoparticles can introduce specific properties to the composites such as superparamagnetism, high or low refractive index, UV absorption, or simply color, depending on the constitution of the particles. However, it has to be considered that materials properties can change in an inefficient way when the particles become too small, as was demonstrated on the example of the refractive index of PbS. In such cases, particle sizes below a certain limit may diminish an attempted effect.

ACKNOWLEDGMENTS

I cordially pronounce my innumerable thanks to my colleagues for their invaluable support to the topics dealt with in this article, in particular (in alphabetical order) H.-J. Althaus, M. Büchler, C. Bastiaansen, C. Darribère, F. H. Dalla Torre, Y. Dirix, M. Gianini, W. Heffels, H. Kiess, T. Kyprianidou-Leodidou, J. F. Löffler, P. Margraf,

M. Müller, R. Nussbaumer, S. E. J. Rigby, M. Russo, J. Schällibaum, P. Smith, N. Stingelin, U. W. Suter, T. Tervoort, D. Vetter, P. Walther, E. Wehrli, M. Weibel, Y. Wyser, and L. Zimmermann.

REFERENCES

[1] D. L. Leslie-Pelecky, R. D. Rieke, *Chem. Mater.* **8**, 1770 (1996).

[2] R. H. Kodama, *J. Magn. Magn. Mater.* **200**, 359 (1999).

[3] J. F. Löffler, H. B. Braun, W. Wagner, *Phys. Rev. Lett.* **85**, 1990 (2000).

[4] U. Kreibig, M. Vollmer, *Optical Properties of Metal Clusters*, Springer, Berlin 1995.

[5] A. J. Kinloch, *Adhesion and Adhesives*, Chapman and Hall, London 1987.

[6] H. Salmang, H. Schloze, *Keramik*, part 1, Springer, Berlin 1982.

[7] A. Carre, J. Vial, in "Preprints EURADH '92", ed. by J. Schultz, W. Brockmann, DECHEMA, Frankfurt am Main 1992, p. 90

[8] G. Carotenuto, B. Martorana, P. Perlo, L. Nicolais, *J. Mater. Chem.* **13**, 2917 (2003).

[9] F. Faupel, V. Zaporojtchenko, H. Greve, U. Schürmann, V.S.K. Chakravadhanula, C. Hanisch, A. Kulkarni, A. Gerber, E. Quandt, R. Podschun, *Contrib. Plasma Phys.* **47**, 537 (2007).

[10] D. J. Barber and I. C. Freestone, *Archaeometry*, 1990, **32**, 33.

[11] A. Neri, *L'arte vetraria*, Giunti, Firenze 1612.

[12] E. D. Palik, Ed. *Handbook of Optical Constants of Solids*, Academic Press, Orlando (Florida) 1985.

[13] Lüdersdorff, *Verhandl. Verein. Beförderung Gewerbefleiss. Preussen* **12**, 224 (1833).

[14] Luedersdorff, *Neuest. Forsch. Techn. Ökonom. Chem.* **3**, 212 (1833); *J. Techn. Ökonom. Chem.* **18**, 212 (1833).

[15] G. Frens, *Colloid Z. Z. Polym.* **250**, 736 (1972).

[16] W. Heffels, J. Friedrich, C. Darribère, J. Teisen, K. Interewicz, C. Bastiaansen, W. Caseri, P. Smith, *Recent Res. Devel., Macromol. Res.* **2**, 143 (1997).

[17] C. Bastiaansen, W. Caseri, C. Darribère, S. Dellsperger, W. Heffels, A. Montali, C. Sarwa, P. Smith, C. Weder, *Chimia* **52**, 591 (1998).

[18] W. Steubing, *Ann. Phys. 4. Folge (Drude's Ann.)* **26**, 329 (1908).

[19] G. Mie, *Phys. Z.* **8**, 769 (1907).

[20] G. Mie, *Ber. Deutsch. Chem. Ges.* **40**, 492 (1907).

[21] Y. Dirix, C. Bastiaansen, W. Caseri, P. Smith, *Adv. Mater.* **11**, 223 (1999).

[22] Y. Dirix, C. Darribère, W. Heffels, C. Bastiaansen, W. Caseri, P. Smith, *Appl. Opt.* **38**, 6581 (1999).

[23] A. Pucci, M. Bernabò, P. Elvati, L. I. Meza, F. Galembeck, C. A. de Paula Leite, N. Tirelli, G. Ruggeri, *J. Mater. Chem.* **16**, 1058 (2006).

[24] H. Ambronn, *Ber. Kgl. Sächs. Ges. d. Wiss. Leipzig, Math.-Phys. Klasse* **48**, 613 (1896).

[25] S. Apáthy, *Mitt. Zool. Stat. Neapel* **12**, 495 (1897).

[26] H. Ambronn, R. Zsigmondy, *Ber. Kgl. Sächs. Ges. d. Wiss. Leipzig, Math.-Phys. Klasse* **51**, 13 (1899).

[27] L. Zimmermann, M. Weibel, W. Caseri, U. W. Suter, P. Walther, *Polym. Adv. Technol.* **4**, 1 (1993).

[28] J. P. Chen, C. M. Sorensen, K. J. Klabunde, G. C. Hadjipanayis, *J. Appl. Phys.* **76**, 6316 (1994).

[29] H. T. Yang, Y. K. Su, C. M. Shen, T. Z. Yang, H. J. Gao, *Surf. Interface Anal.* **36**, 155 (2004).

[30] V. F. Puntes, K. M. Krishnan, A. P. Alivisatos, *Science* **291**, 2115 (2001).

[31] Y. Bao, B. Pakhomov, K. M. Krishnan, *J. Appl. Phys.* **97**, 10 J317 (2005).

[32] D. P. Dinega, M. G. Bawendi, *Angew. Chem. Int. Ed.* **38**, 1788 (1999).

[33] J. Schällibaum, F. H. Dalla Tore, W. R. Caseri, J. F. Löffler, *Nanoscale* **1**, 374 (2009).

[34] H. Shao, Y. Huang, *J. Appl. Phys.* **99**, 08N702 (2006).

[35] N. Shukla, E. B. Svedberg, J. Ell, A. J. Roy, *Mater. Lett.* **60**, 1950 (2006).

[36] S. Sun, C. B. Murray, *J. Appl. Phys.* **85**, 4325 (1999).

[37] N. Wu, L. Fu, M. Su, M. Aslam, K. C. Wong, V. P. Dravid, *Nano. Lett* **4**, 383 (2004).

[38] J. F. Ciebien, R. E. Cohen, A. Duran, *Supramol. Sci.* **5**, 31 (1998).

[39] H. B. Sunkara, J. M. Jethmalani, W. T. Ford, *Chem. Mater.* **6**, 362 (1994).

[40] J. M. Jethmalani, W. T. Ford, *Chem. Mater.* **8**, 2138 (1996).

[41] M. Gianini, W. R. Caseri, U. W. Suter, *J. Phys. Chem. B* **105**, 7399 (2001).

[42] C. Yee, M. Scotti, A. Ulman, H. White, M. Rafailovich, J. Sokolov, *Langmuir* **15**, 4314 (1999).

[43] Y.-S. Shon, S. M. Gross, B. Dawson, M. Porter, R. W. Murray, *Langmuir* **16**, 6555 (2000).

[44] T. Kunitake, in *Physical Chemistry of Biological Interfaces*, ed. by A. Baszkin, W. Norde, Marcel Dekker, New York 2000, p 283.

[45] D. Needham, D. Zhelev, in *Giant Vesicles*, ed. by P. L. Luisi, P. Walde, John Wiley & Sons, Chichester 2000, p 103.

[46] J. M. Seddon, R. H. Templer, in *Structure and Dynamics of Membranes*, Vol. 1, ed. by R. Lipowsky, E. Sackmann, Elsevier, Amsterdam 1995, p 97.

[47] E. Sackmann, in *Structure and Dynamics of Membranes*; Vol. 1, ed. by R. Lipowsky, E. Sackmann, Elsevier, Amsterdam 1995, p 213.

[48] U. Seifert, R. Lipowsky, in *Structure and Dynamics of Membranes*, Vol. 1, ed. by R. Lipowsky, E. Sackmann, Elsevier, Amsterdam 1995, p. 403.

[49] C. K. Chan, J. F. Porter, Y.-G. Li, W. Guo, C.-M. Chan, *J. Am. Ceram. Soc.* **82**, 566 (1999).

[50] P. Christensen, A. Dilks, T. A. Egerton, J. Temperley, *J. Mater. Sci.* **35**, 5353 (2000).

[51] U. Gesenhues, *Polym. Degrad. Stab.* **68**, 185 (2000).

[52] H. F. Mark, J. McKetta, and D. Othmer, (eds.), *Kirk-Othmer Encyclopedia of Chemical Technology*, 2nd Edition, Vol. 20, John Wiley and Sons, New York 1969.

[53] R. J. Nussbaumer, P. Smith, W. Caseri, *J. Nanosci. Nanotechnol.* **7**, 2422 (2007).

[54] R. J. Nussbaumer, W. R. Caseri, P. Smith, T. Tervoort, *Macromol. Mater. Eng.* **288**, 44 (2003).

[55] J. C. Seferis, in *Polymer Handbook*, 3rd edition, ed. by J. Brandrup, E. H. Immergut, John Wiley & Sons, New York 1989, p. VI/451.

[56] J. C. Seferis, R. J. Samuels, *Polym. Eng. Sci.* **19**, 975 (1979).

[57] A. R. Wedgewood, J. C. Seferis, *Polym. Eng. Sci.* **24**, 328 (1984).

[58] F. Papadimitrakopoulos, P. Wisniecki, D. E. Bhagwagar, *Chem. Mater.* **9**, 2928 (1997).

[59] C. Renz, *Helv. Chim. Acta* **4**, 961 (1921).

[60] R. P. Müller, J. Steinle, H. P. Boehm, *Z. Naturforsch. B* **45**, 864 (1990).

[61] B. Ohtani, S. Adzuma, S.-I. Nishimoto and T. Kagiya, *J. Polym. Sci. C* **25**, 383 (1987).

[62] R. J. Nussbaumer, W. R. Caseri, P. Smith, *J. Nanosci. Nanotechnol.* **6**, 459 (2006).

[63] M. Russo, S. E. J. Rigby, W. Caseri, N. Stingelin, *J. Mater. Chem.* **20**, 1348 (2010).

[64] A. Ammala, A. J. Hill, P. Meakin, S. J. Pas, T. W. Turney, *J. Nanopart. Res.* **4**, 167 (2002).

[65] Y.-Q. Li, S.-Y. Fu, Y.-W. Mai, *Polymer* **47**, 2127 (2006).

[66] M. M. Demir, K. Koynov, Ü. Akbey, C. Bubeck, I. Park, I. Lieberwirth, G. Wegner, *Macromolecules* **40**, 1089 (2007).

[67] H. Althues, P. Simon, F. Philipp, S. Kaskel, *J. Nanosci. Nanotechnol.* **6**, 409 (2006).

[68] T. Kyprianidou-Leodidou, P. Margraf, W. Caseri, U. W. Suter, P. Walther, P. *Polym. Adv. Technol.* **8**, 505 (1997).

[69] R. J. Herold, *Can. J. Chem.* **41**, 1368 (1963).

[70] M. Weibel, W. Caseri, U. W. Suter, H. Kiess, E. Wehrli, *Polym. Adv. Technol.* **2**, 75 (1991).

[71] F. T. Burling, B. M. Goldstein, *J. Am. Chem. Soc.* **114**, 2313 (1992).

[72] L. Zimmermann, M. Weibel, W. Caseri, U. W. Suter, *J. Mater. Res.* **8**, 1742 (1993).

[73] C.-M. Chang, C.-L. Chang, C.-C. Chang, *Macromol. Mater. Eng.* **291**, 1521 (2006).

[74] C. Lü, Y. Cheng, Y. Liu, F., Liu, B. Yang, *Adv. Mater.* **18**, 1188 (2006).

[75] J. Brandrup, E. H. Immergut (eds.), *Polymer Handbook*, John Wiley and Sons, New York. 1989.

[76] T. Tsuzuki, T., *Macromol. Mater. Eng.* **293**, 109 (2008).

[77] T. Kyprianidou-Leodidou, W. Caseri, U. W. Suter, *J. Phys. Chem.* **98**, 8992 (1994).

[78] T. Kyprianidou-Leodidou, H.-J. Althaus, Y. Wyser, D. Vetter, M. Büchler, W. Caseri, U. W. Suter, *J. Mater. Res.* **12**, 2198 (1997).

[79] R. A. Berner, *Am. J. Sci.* **265**, 773 (1967).

[80] D. T. Rickard, *Am. J. Sci.* **275**, 636 (1975).

[81] D. J. Vaughan, J. Craig, *Mineral Chemistry of Metal Sulfides*, Cambridge University Press 1978.

[82] P. Taylor, *Am. Mineral.* **65**, 1026 (1980).

3

PREPARATION AND CHARACTERIZATION OF METAL–POLYMER NANOCOMPOSITES

L. Nicolais[1] and G. Carotenuto[2]

[1]*Dipartimento di Ingegneria dei Materiali e della Produzione, Università "Federico II" di Napoli, Napoli, Italy*
[2]*Institute for Composite and Biomedical Materials, National Research Council, Napoli, Italy*

3.1 INTRODUCTION

The availability of new materials is the main requirement for a rapid expansion of different technological fields. Materials with finely tunable properties, unusual property combinations (e.g., materials that combine optical transparency to magnetism or fluorescence), and new physical–chemical properties are strictly required. Nano-sized metals represent a novel material class that may solve a number of present and future technological problems because of their unique characteristics that can be advantageously exploited for applications in different technological fields. However, the difficult handling of such extremely small structures (particles with diameters of only a few nanometers) represents a strong limitation to their use. At present, very few approaches are available for building functional devices using metal nanoparticles (e.g., manipulations of single nanoparticles by surface tunneling microscopy, spontaneous self-assembly, di-electrophoresis). In addition, most of metal nanoparticles are not thermodynamically stable since they can aggregate because of the high surface free-energy content, be

Nanocomposites: In Situ *Synthesis of Polymer-Embedded Nanostructures*, First Edition.
Edited by Luigi Nicolais and Gianfranco Carotenuto.
© 2014 John Wiley & Sons, Inc. Published 2014 by John Wiley & Sons, Inc.

oxidized by air oxygen, and be contaminated by absorption of SO_2 and other small molecules. The use of dielectric polymers as embedding phase for the nanoscopic metal particles represents a valid solution for these manipulation–stabilization problems. This solution is very convenient from a technological point of view, since depending on their structure, polymers show a large variety of physical characteristics (e.g., they can be an electrical/thermal insulator or conductor, may have a hydrophobic or hydrophilic character, can be mechanically rigid, plastic, or rubbery); in addition, polymer-embedded metal nanoparticles can be widely processed by injection molding, extrusion, etc. Finally, polymer embedding represents an easy and convenient approach to metal nanoparticle stabilization, which allows their handling and technological use. The material resulting from nanoparticle embedding in polymers is frequently indicated as nanocomposites because of the biphasic nature.

The development of metal–polymer nanocomposites looks back at a long history that is connected to the names of some famous scientists. The oldest technique for the preparation of metal–polymer nanocomposites is described in an abstract that appeared in 1835. In an aqueous solution, a gold salt was reduced in the presence of gum arabic, and subsequently a nanocomposite material was obtained in the form of a purple solid simply by coprecipitation with ethanol. Around 1900, widely forgotten reports notify the preparation of polymer nanocomposites with uniaxially oriented inorganic particles and their remarkable optical properties. Dichroic plant and animal fibrils (e.g., linen, cotton, spruce, or chitin, among others) were prepared by impregnation with solutions of silver nitrate, silver acetate, or gold chloride, followed by reduction of the corresponding metal ions under the action of light. Dichroic films were also obtained using gold chloride-treated gelatin that was subsequently drawn, dried, and finally exposed to light. Similar results were obtained when gelatin was mixed with colloidal gold before drying and drawing. In 1904, Zsigmondy (Nobel Laureate in Chemistry, 1925) reported that nanocomposites of colloidal gold and gelatin reversibly changed the color from blue to red upon swelling with water. In order to explain the mechanism of nanocomposite color change, they suggested that the material absorption must also be influenced by the interparticle distance. In addition, around the same time, the colors of gold particles embedded in dielectric matrices were subject of detailed theoretical analyses by Maxwell Garnett who explained the color shifts upon variation of particle size and volume fraction in a medium. During the following three decades, dichroic fibers were prepared with many different metals (i.e., Pd, Pt, Cu, Ag, Au, Hg). The dichroism was found to depend strongly on the employed element, and optical spectra of dichroic nanocomposites, made of stretched poly(vinyl alcohol) films containing gold, silver, or mercury, were presented in 1946. It was assumed already in the early reports that dichroism was originated by the linear arrangement of small particles or by polycrystalline rodlike particles located in the uniaxially oriented spaces present in the fibers. An electron micrograph depicted in 1951 showed that tellurium needles were present inside a dichroic film made of stretched poly(vinyl alcohol). In 1910, Kolbe proved that dichroic nanocomposite samples based on gold contained the metal indeed in its zero-valence state. Such affirmation was confirmed a few years later by X-ray powder diffraction (XRD); in particular it was shown that zero-valence silver and gold were present in the respective nanocomposites made with oriented ramie fibers, and the ringlike interference patterns of the metal crystallites showed that the individual primary crystallites were not oriented. Based on the Scherrer formula, which was developed just in this period, the average particle diameter of silver and gold crystallites was determined in fibers of ramie, hemp, bamboo, silk, wool, viscose, and cellulose acetate to be between 5 nm and 14 nm.

Metals undergo a considerable property change by size reduction, and their composites with polymers are very simple systems with interesting functional properties. The new properties observed in nano-sized metals (mesoscopic metals) are produced by quantum-size effects (i.e., electron confinement and surface effect). These properties are size dependent and can be simply tuned by changing the particle size. Since the same element may show different sets of properties by size variation, a 3D periodic table of elements has been proposed. Confinement effects arise in nano-sized metal domains since conduction electrons are allowed to move in a very small space, which is comparable to their de Broglie wavelength; consequently their states are quantized just like in the atoms, and these systems are termed "artificial atoms." Surface effects are produced because with a decrease in size, matter constitutes more and more of surface atoms instead of inner atoms. As a result, the matter properties slowly switch from that determined by the characteristics of inner atoms to that belonging to surface atoms. In addition, the surface nature of a nano-sized object significantly differs from that of a massive object. Atoms on the surface of a massive object are principally located on basal planes, but they transform almost completely in edge and corner atoms with a decrease in size. Because of the very low coordination number, edge and corner atoms are highly chemically reactive, super-catalytically active, highly polarizable, etc., in comparison with atoms on basal planes.

Because of quantum-size effects, mesoscopic metals show a set of properties completely different from that of their massive counterpart. Particularly interesting are the size-dependent ferromagnetism and the superparamagnetism characterizing all metals (included diamagnetic metals like silver); the chromatism observed with silver, gold, and copper metals due to plasmon absorption; the photo- and thermal luminescence; and the super-catalytic effect (hyperfine catalysts are characterized by an extraordinarily higher catalytic activity and a different selectivity compared to corresponding fine powders). In addition, because of the band-structure disappearance, metals become thermal and electrical insulators at very small sizes. They are highly chemically reactive (heterogeneous reactions become stoichiometric and new reaction schemes are possible, e.g., nano-sized noble metals are quite reactive) and strongly absorbent and show completely different thermodynamic properties (e.g., nanoscopic metals melt at much lower temperatures). Many of these unique chemical–physical characteristics of nano-sized metals leave unmodified after embedding in polymers (e.g., optical, magnetic, dielectric, and thermal-transport properties), and therefore they can be used as polymer additives that provide plastic materials of special functionalities.

A limited number of methods have been developed for the preparation of metal–polymer nanocomposites. Usually, such techniques consist of highly specific approaches, which can be classified as *in situ* and *ex situ* methods. In the *in situ* methods, two steps are needed: firstly, the monomer is polymerized in solution, with metal ions introduced before or after polymerization. Then metal ions in the polymer matrix are reduced chemically, thermally, or by ultraviolet (UV)/γ irradiation. In the *ex situ* processes, the metal nanoparticles are chemically synthesized, and their surface is organically passivated. The derivatized nanoparticles are dispersed into a polymer solution or liquid monomer that is then polymerized.

For the comprehension of mechanisms involved in the appearance of novel properties in polymer-embedded metal nanostructures, their characterization represents the fundamental starting point. The microstructural characterization of nanofillers and nanocomposite materials is performed mainly by transmission electron microscopy (TEM), large-angle XRD, and optical spectroscopy (UV–visible (Vis)). These three techniques are very effective to determine particle morphology, crystal structure, composition, and grain size.

Of the many techniques that have been used to study the structure of metal–polymer nanocomposites, TEM has undoubtedly been the most useful. This technique is currently used to probe the internal morphology of nanocomposites. High-quality images can be obtained because of the presence in the sample of regions that do not allow high-voltage electron beam passage (i.e., the metallic domains) and region perfectly transparent to the electron beam (i.e., the polymeric matrix). High-resolution transmission electron microscopy (HRTEM) allows morphological investigations with resolution of 0.1nm, and thus this technique makes possible to accurately image nanoparticle sizes, shape, and atomic lattice.

Large-angle X-ray powder diffraction (XRD) has been one of the most versatile techniques utilized for the structural characterization of nanocrystalline metal powders. The modern improvements in electronics, computers, and X-ray sources have allowed it to become an indispensable tool for identifying nanocrystalline phases as well as crystal size and crystal strain. The comparison of the crystallite size obtained by the XRD diffractogram using the Scherrer formula with the grain size obtained from the TEM image allows to establish if the nanoparticles have a mono- or polycrystalline nature.

Metal clusters are characterized by the surface plasmon resonance, which is an oscillation of the surface plasma electrons induced by the electromagnetic field; consequently their microstructure can be indirectly investigated by optical spectroscopy (UV–Vis spectroscopy). The characteristics of this absorption (shape, intensity, position, etc.) are strictly related to the nature, structure, topology, etc., of the cluster system. In fact, the absorption frequency is a fingerprint of the particular metal, the eventual peak splitting reflects aggregation phenomena, the intensity of the peak is related to the particle size, the absorption wavelength is related to the particle shape, the shift of the absorption with increasing of temperature is indicative of a cluster melting, etc. For bimetallic particles, information about inner structure (intermetallic or core/shell) and composition can be obtained from the maximum absorption frequency. Differently from *off-line* techniques (e.g., TEM, XRD), this method allows *on-line* and *in situ* cluster sizing and monitoring of morphological evolution of the system. This method has been used also in the study of cluster nucleation and growth mechanisms.

Advances in Raman spectroscopy, energy-dispersive spectroscopy, infrared spectroscopy, and many other techniques are of considerable importance as well. In fact, the success that nanostructured materials are having in the last few years is strictly related to the advanced characterization techniques that are today available.

Applications of metal–polymer nanocomposites have already been made in different technological fields; however, the use of a much larger number of devices based on these materials can be predicted for the next future.

Because of the plasmon surface absorption band, atomic clusters of metals can be used as pigments for optical plastics. The color of the resulting nanocomposites is light-fast and intensive; in addition, these materials are perfectly transparent, since the cluster size is much lower than light wavelength. Gold, silver, and copper can be used for color filter application. Also UV absorbers can be made, for example, by using Bi clusters. The plasmon surface absorption frequency is modulated by making intermetallic particles (e.g., Pd/Ag, Au/Ag) of adequate composition.

Polymeric films containing uniaxially oriented pearl-necklace type of nanoparticles arrays exhibit a polarization-dependent and tunable color. The color of these systems is very bright and can change strongly modifying the light polarization direction. These materials are obtained by dispersing metal nanoparticles in polymeric thin films and subsequently reorganizing the dispersed phase into pearl-necklace arrays by solid-state drawing at temperature below the polymer melting point. The formation of these arrays in the films is the cause of a strong polarization-direction-dependent color, which can be used in the fabrication of liquid-crystal color display and special electro-optical devices.

Surface plasmon resonance has been used to produce a wide variety of optical sensors, that is, systems that are able to change their color in presence of specific analytes. These devices can be used as sensors for immunoassay, gas, and liquid.

Metals are characterized by ultrahigh/infralow refractive indices and therefore can be used to modify the refractive index of optical plastics. Ultrahigh/infralow refractive index optical nanocomposites can be used in the waveguide technology (e.g., planar waveguides and optical fibers).

Plastics doped by atomic clusters of ferromagnetic metals show magneto-optical properties (i.e., when subject to a strong magnetic field, they can rotate the vibration plane of a plane-polarized light), and therefore they can be used as Faraday rotators. These devices have a number of important optical applications (e.g., magneto-optical modulators, optical isolators, optical shutters).

Nano-sized metals (e.g., gold, silver) have attracted much interest because of the nonlinear optical polarizability, which is caused by the quantum confinement of the metal electron cloud. When irradiated with light above a certain threshold power, the optical polarizability deviates from the usual linear dependence on that power. By incorporating these particles into a clear polymeric matrix, nonlinear optical devices can be made in a readily processable form. These materials are used to prepare a number of devices for photonics and electro-optics.

Finally, polymer embedding represents a simple but effective way to use mesoscopic properties of nano-sized metals. A large variety of advanced functional devices can be based on this simple material class. In the last few years, a number of pioneering techniques have been developed for preparing metal–polymer nanocomposite materials. In particular, the *in situ* techniques based on the thermolysis of special organic metal precursors seem to be a very promising approach, principally for the possibility to produce metal–polymer nanocomposites on a large scale by techniques already available for thermoplastic polymer hot processing.

3.2 NANOCOMPOSITE PREPARATION

A number of noble metals, semimetals, and metal sulfides can be easily generated by thermal decomposition of mercaptides. Mercaptides are sulfur-based organic compounds, formally corresponding to thiol salts. Mercaptides are also referred to as thiolates and linear alkane-thiolates (i.e., $MeSC_nH_{2n+1}$) represent the most common mercaptide class.

Depending on the mercaptide type and precisely on the nature of the metal–sulfur bond, the inorganic product of thermolysis can be the pure element or its binary compounds with sulfur. In particular, covalent mercaptides (e.g., mercaptides of noble metals and semimetals) lead to a pure zero-valence solid phase, while ionic mercaptides (e.g., mercaptides of IV-period transition metals) give the metal sulfide (i.e., Me_xS_y) as thermolysis product. There are also a few examples of mercaptides whose thermolysis gives a mixture of metal and sulfide as decomposition product (e.g., antimony mercaptide, $Sb(SR)_3$, and silver mercaptide, $AgSR$). The mechanisms involved in these two thermolysis reactions are quite different (see Fig. 3.1). When thermolysis produces a metal or a semimetal phase, the reaction mechanism is based on the homolytic dissociation of the metal–sulfur bonds with formation of $RS \cdot$ radicals, which combine together leading to disulfide molecules (RSSR). Instead, ionic mercaptides are dissociated in ions at molten state, and consequently a nucleophilic substitution (referred to as SN_2

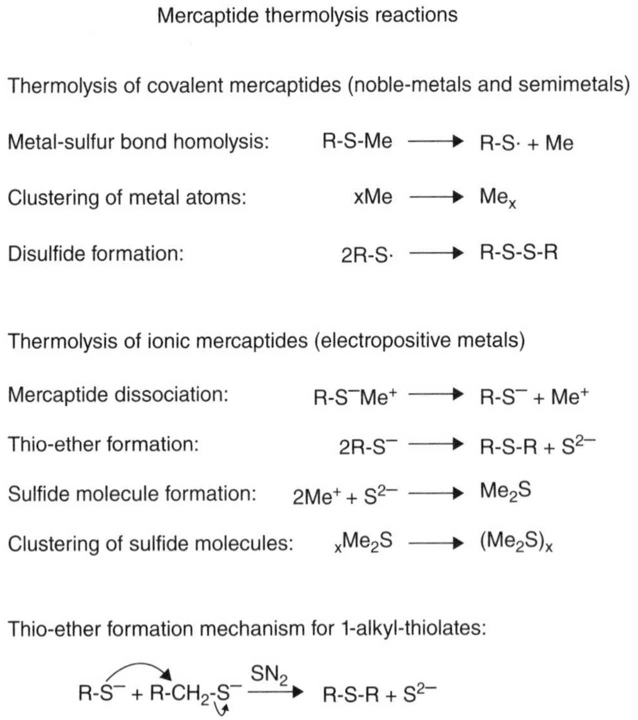

Figure 3.1. Mechanistic pathways of mercaptide thermal decomposition.

pathway), involving the thiolate group, RS^-, and the α-carbon of another thiolate molecule, takes place during the thermal treatment giving the formation of metal sulfide and thioether (RSR). Thiolate molecules, RS^-, are strong nucleophiles, and also the sulfur ion, S^{2-}, is a good leaving group. Examples of these two reaction schemes are the thermolysis of the following mercaptides:

$$2\,Bi(SR)_3 \rightarrow 2\,Bi + 3\,RSSR \qquad (3.1)$$

$$Pb(SR)_2 \rightarrow PbS + RSR \qquad (3.2)$$

Thus, the reaction mechanism is influenced by the nature of the alkyl group, R-, since mercaptides of secondary and tertiary thiols and thiophenols cannot give nucleophilic substitution (SN_2). Thiophenol thermolysis usually gives the metal sulfide with diphenyl molecule formation as by-product. However, these ionic compounds melt and decompose at very high temperatures, and therefore they are not adequate for the nanocomposite synthesis.

Mercaptide thermolysis takes place at quite moderate temperatures ($150°–250°C$). Such mild thermal conditions are absolutely compatible with thermal stability of common polymers, and consequently the thermal degradation of mercaptide molecules can be also performed with mercaptide dissolved into a polymeric medium. In this case, a finely dispersed inorganic solid phase, embedded in polymer, is generated. Materials based on clusters confined in polymeric matrices are named nanocomposites [1–4]. Both semiconductor–polymer and metal–polymer nanocomposites have unique functional properties that can be exploited for applications in several advanced technological fields (e.g., optics, nonlinear optics, magneto-optics, photonics, optoelectronics) [1].

According to the described reaction schemes, in addition to the inorganic phase, the mercaptide thermal decomposition produces also disulfide or thioether molecules as organic by-product. In some cases, these molecules are strongly chemisorbed on the surface of the produced metals (e.g., gold, palladium, platinum, silver), but also in the case of a sulfide product, they can be physically adsorbed, leading to polymer-embedded cluster compounds (i.e., $Me_x(SR)_y$, with $y \ll x$) at reaction end. In other cases, the RSR or RSSR molecules are dissolved in the polymer matrix, causing a light plasticizing effect (typically, a lowering in the polymer glass-transition temperature is observed). At very high temperatures ($300°–400°C$), the disulfide molecules convert to thioether and polysulfides, according to the following reaction scheme:

$$2\,R\text{-}S\text{-}S\text{-}R \rightarrow R\text{-}S\text{-}R + R\text{-}S\text{-}S\text{-}S\text{-}R \qquad (3.3)$$

The mercaptide thermolysis reaction is well known in organic chemistry, and it may represent a chemical route for thioether syntheses. However, the aforementioned reaction schemes have been only recently considered for the generation of inorganic solids. In particular, the use of mercaptides to produce semiconductor sulfide films on ceramic substrates has been proposed in the literature [5]. Mercaptides of gold and silver have been also used to make special inks for the metallization of pottery and ceramic substrates [6–11]. More recently, mercaptides have been used to generate finely dispersed inorganic phases inside a polymeric matrix (nanocomposites) [12–14], and a "solventless" approach for the production of monodispersed nanostructures has been also proposed [15–17].

Usually, covalent mercaptides are characterized by a high solubility in nonpolar organic media like ethers, hydrocarbons, and chlorurate hydrocarbons; consequently these compounds may dissolve into hydrophobic polymers (e.g., polystyrene, poly(methyl methacrylate), polycarbonate, poly(vinyl acetate)). Mercaptide/polymer blends are prepared by solution-casting technology (to prevent mercaptide decomposition before the thermal annealing treatment). Both mercaptide and polymer are dissolved into an organic solvent, then this mixture is cast on a glassy plate (e.g., Petri dish), and the solvent is slowly allowed to evaporate in air at room temperature. A slow solvent removal is required to avoid mixture cooling with mercaptide precipitation. Contact-free dispersions of particles in polymer result from the thermal annealing of homogeneous mercaptide/polymer blends. In addition, this approach gives very small (from a few nanometers to a few nanometer tens) and monodispersed clusters. The solubility of ionic mercaptides into hydrophobic polymers is quite low, but it can be improved by increasing the size of the alkyl group (–R). Usually, covalent dodecyl-mercaptides (e.g., $AgSC_{12}H_{25}$) are moderately soluble in polystyrene, while octadecyl-mercaptides (e.g., $AgSC_{18}H_{37}$) are quite soluble.

The scarce solubility of mercaptide molecules in most organic media is related to their polymeric nature. In fact, sulfur bridges among the metal atoms are frequently present in the crystalline structure (lamellar structure) of these sulfur-based compounds. Such polymeric structures can be completely destroyed by treatment with strong ligand molecules like N-methyl-imidazole, which coordinate to the metal atoms, preventing the formation of sulfur bridges (i.e., $[Me(SR)_2]_x + nxL \rightarrow xMe(SR)_2L_n$). Heteroleptic mercaptides are much more soluble than the corresponding homoleptic compounds, and their thermolysis takes place in two steps—(i) imidazole ligand lost and (ii) metal–sulfur bond cleavage at a higher temperature:

$$Me(SR)_2L_n \rightarrow nL + Me(SR)_2 \rightarrow Me + (SR)_2 \qquad (3.4)$$

For a controlled and uniform heating of the mercaptide/polymer blends, samples are shaped in the form of films that are annealed on a hot plate at temperatures ranging from 150 °C to 200 °C. Both film surfaces must be simultaneously heated, and therefore they are placed between two heated metallic surfaces. Thermal gradients on the heating surface must be limited by interposing a large metallic block (heat reservoir) between the sample and heat source. Thermal annealing of mercaptide/polymer blends can be also performed by using infrared lamps or a low-energy laser spot (ca. 100 mW).

3.3 MERCAPTIDE SYNTHESIS

Most mercaptides are not commercially available products, since large-scale applications of these chemical compounds are quite limited (only mercaptides of tin and antimony are industrially used as thermal stabilizers for poly(vinyl chloride) [18, 19]); however, these compounds can be synthesized in a very simple way. Owing to their low water solubility, mercaptides can precipitate by reacting thiols (or thiophenols) with aqueous solutions of the corresponding metal salts. Mercaptides of mercury, lead, zinc, and copper are well-known substances, and many others have been prepared (e.g., mercaptides of silver, gold, platinum, palladium, iridium, nickel, iron, cobalt, antimony, bismuth, cadmium):

$$(CH_3COO)_2Pb + 2RSH \rightarrow Pb(SR)_2 + 2CH_3COOH \qquad (3.5)$$

A high reaction yield characterizes the synthesis of most mercaptides and it is related to the nonpolar mercaptides' nature. Some mercaptides have a so low solubility in water and alcohols that promptly precipitate from these media; also in the case, a strong acid is simultaneously generated (e.g., when thiols are reacted with silver nitrate). However, to achieve complete mercaptide precipitation, the acid by-product should be neutralized, or the reaction should be performed using metal salts of weak acids (like mercury cyanide, mercury oxide, lead acetate).

Owing to the reductant nature of thiols, a change in the oxidation number of metallic ions can be observed before mercaptide precipitation (e.g., Au(III) ions are reduced to Au(I) before mercaptide precipitation). In addition, in the case of polyvalent metals, intermediate reaction products can be obtained, since the mercaptide formation takes place by steps. For example, chlorides of alkylmercapto-mercury can be obtained from the reaction of thiols with mercury chloride:

$$HgCl_2 + RSH \rightarrow RSHgCl + HCl \qquad (3.6)$$

Mercaptides should be stored in a dry environment (desiccator cabinet) to avoid hydrolysis with thiol and metal hydroxide formation. In addition, some mercaptides can be oxidized in air, for example, lead mercaptides of high-molecular-weight thiols are soluble in hydrocarbons but lead to insoluble peroxides by reaction with atmospheric oxygen. Mercaptide of heavy metals give exchange reaction, and in some cases it may represent a useful synthetic route:

$$MeX + NaSR \rightarrow MeSR + NaX \qquad (3.7)$$

Since thiols are acid compounds (hydrosulfuric acid derivatives), mercaptides can be also obtained by reacting them with organometallic compounds [20, 21]:

$$Sb(CH_2R')_3 + 3R\text{-}SH+ \rightarrow Sb(SR)_3 + 3CH_3\text{-}R' \qquad (3.8)$$

The preparation of nanocomposite materials by thermolysis of mercaptide molecules dissolved in polymer represents a quite universal approach, and this reaction scheme is only limited by the ability to synthesize the mercaptide precursor. A short description of most common mercaptide preparation follows:

Bismuth Mercaptide. Bismuth(II) dodecyl-mercaptide, $Bi(SC_{12}H_{25})_3$, was synthesized by reacting stoichiometric amounts of dodecanethiol ($C_{12}H_{25}SH$, Aldrich) and bismuth(III) chloride ($BiCl_3$, Aldrich). Both reactants were dissolved in ethyl alcohol (99.8%, Fluka) at room temperature, and these solutions were mixed together under stirring. The presence of a little amount of water in ethyl alcohol caused the formation of bismuth hydroxide ($Bi(OH)_3$). Consequently, a few drops of HCl solution were added to the alcoholic bismuth salt solution in order to dissolve this hydroxide. Mercaptide precipitation did not take place just after the reactant mixing, but the addition of ammonium hydroxide ($NH_3 \cdot H_2O$, Aldrich) to neutralize the equilibrium HCl was required. Bismuth dodecyl-mercaptide was a waxy solid of yellow color, characterized by a melting point of 64 °C. The mercaptide was isolated by vacuum filtration and purified by dissolution/precipitation from chloroform/ethyl alcohol. $Bi(SC_{12}H_{25})_3$ thermolysis took place at a very low temperature (100 °C) and gave pure zero-valence bismuth.

Antimony Mercaptide. Antimony(III) dodecyl-mercaptide, $Sb(SC_{12}H_{25})_3$, was prepared by adding drop by drop an alcoholic solution of dodecanethiol ($C_{12}H_{25}SH$, Aldrich) to an antimony chloride solution ($SbCl_3$, Aldrich, 99.9%) in ethanol (99.8%, Fluka) at room temperature, under stirring. Stoichiometric amounts of reactants were used. Mercaptide precipitation did not take place just after reactant mixing, but the addition of ammonium hydroxide ($NH_3 \cdot H_2O$, Aldrich) to neutralize the equilibrium HCl was required. A white crystalline powder promptly precipitated. Such mercaptide powder was separated by vacuum filtration and then accurately washed with ethanol. The mercaptide was purified from NH_4Cl traces by dissolution in chloroform followed by filtration and solvent evaporation. The melting point of $Sb(SC_{12}H_{25})_3$ was of 48 °C, and its thermolysis gave a mixture of Sb_2S_3 (stibnite) and zero-valence antimony (90:10 respectively, as evaluated by XRD).

Silver Mercaptide. Silver(I) dodecyl-mercaptide, $AgSC_{12}H_{25}$, was prepared by adding drop by drop an acetone solution of dodecanethiol ($C_{12}H_{25}SH$, Aldrich) to a silver nitrate soluion ($AgNO_3$, Aldrich, 99.9%) in acetonitrile at room temperature, under stirring [22]. Stoichiometric amounts of unpurified reactants were used. To avoid photochemical decomposition of mercaptide molecules, the reaction vessel was wrapped with an aluminum foil. A white crystalline powder promptly precipitated. The mercaptide powder was separated by pump filtration, washed several times with acetone, and stored in a dry atmosphere. The melting point of $AgSC_{12}H_{25}$ was of 180 °C, and it thermally decomposed at the same temperature, leading to a mixture of zero-valence silver and silver sulfide (Ag_2S). The thermal decomposition of silver dodecyl-mercaptide in polystyrene leaves only the zero-valence silver phase [23].

Platinum Mercaptide. Platinum(II) dodecyl-mercaptide, $Pt(SC_{12}H_{25})_2$, was synthesized by adding a stoichiometric amount of dodecanethiol to an alcoholic solution of Pt(IV) salt ($PtCl_4$, Aldrich). Thiol reduced the platinum(IV) ions to platinum(II) before the mercaptide precipitation. Platinum(II) dodecyl-mercaptide was a crystalline orange solid. The mercaptide was separated by vacuum filtration and accurately washed with ethyl alcohol. The thermal decomposition of $Pt(SC_{12}H_{25})_2$ gave pure zero-valence platinum.

Palladium Mercaptide. Palladium(II) nitrate ($Pd(NO_3)_2$, Aldrich) and dodecane-thiol ($C_{12}H_{25}SH$, Aldrich) were used to synthesize palladium(II) dodecyl-mercaptide (i.e., $Pd(SC_{12}H_{25})_2$). Both reactants were not further purified. The palladium salt was dissolved in heptane, and a thiol solution in heptane was added to it under magnetic stirring. The resulting solution had a strong orange coloration, and mercaptide was completely precipitated by adding some ethanol to this system. The solid precipitate was separated by vacuum filtration and washed with acetone. Then, it was purified by dissolution/precipitation from chloroform/ethanol. The thermal decomposition of $Pd(SC_{12}H_{25})_2$ gave pure zero-valence palladium.

Gold Mercaptide. Gold(I) dodecyl-mercaptide (i.e., $AuSC_{12}H_{25}$) was synthesized by treating an ethanol solution of gold tetra-chloroauric acid ($HAuCl_4 \cdot 3H_2O$, Aldrich) with an ethanol solution of 1-dodecanethiol ($C_{12}H_{25}SH$, Aldrich), at room temperature under stirring. Au(III) was first reduced to Au(I) and then it precipitated as mercaptide ($AuSC_{12}H_{25}$). The obtained light-yellow solid phase was separated by filtration and washed with acetone. The mercaptide can be recrystallized from chloroform, and its thermal decomposition at c. 160 °C gave the pure zero-valence gold.

Iron Mercaptide. Iron(II) dodecyl-mercaptide, $Fe(SC_{12}H_{25})_2$, was synthesized by adding a stoichiometric amount of sodium dodecyl-mercaptide ($NaSC_{12}H_{25}$) to iron(II) chloride ($FeCl_2$, Aldrich). Both reactants were dissolved in distilled water and the reaction was performed at room temperature under stirring. Sodium dodecyl-mercaptide was prepared by neutralizing dodecanethiol ($C_{12}H_{25}SH$, Aldrich) with sodium hydroxide (NaOH, Aldrich) in ethanol. The obtained $Fe(SC_{12}H_{25})_2$ was a microcrystalline powder of green color, which spontaneously decomposes in a few days (disproportion-ation), leading to zero-valence iron and $Fe(SC_{12}H_{25})_3$ of yellow color. Similarly, iron(III) mercaptide can be obtained starting from a Fe(III) salt. The thermal degradation of Fe(II) and Fe(III) mercaptides under nitrogen produces iron sulfide.

Cadmium Mercaptide. Cadmium(II) dodecyl-mercaptide (i.e., $Cd(SC_{12}H_{25})_2$) was prepared by reacting cadmium nitrate (i.e., $Cd(NO_3)_2 \cdot 4H_2O$, Aldrich) with 1-dodecanethiol in ethanol at room temperature under stirring. White microcrys-talline $Cd(SC_{12}H_{25})_2$ promptly precipitated. This mercaptide was soluble in chlo-roform and decomposed at 250 °C, leading to a pure CdS phase.

Zinc Mercaptide. Zinc nitrate ($Zn(NO_3)_2$, Aldrich) was dissolved in ethanol, and an aqueous solution of ammonium hydroxide was added drop by drop until the metal hydroxide was completely dissolved. Then, an alcoholic solution of 1-dodecanethiol was added to this system at room temperature under stirring, and the precipitate was isolated by vacuum filtration. Zn mercaptide (i.e., $Zn(SC_{12}H_{25})_2$) was dissolved in chloroform and blended with polystyrene, leading to translucent films after solvent removal. The thermal annealing of zinc mercaptide/polystyrene blends, performed under nitrogen at 250 °C, gave a nanoscopic ZnS phase.

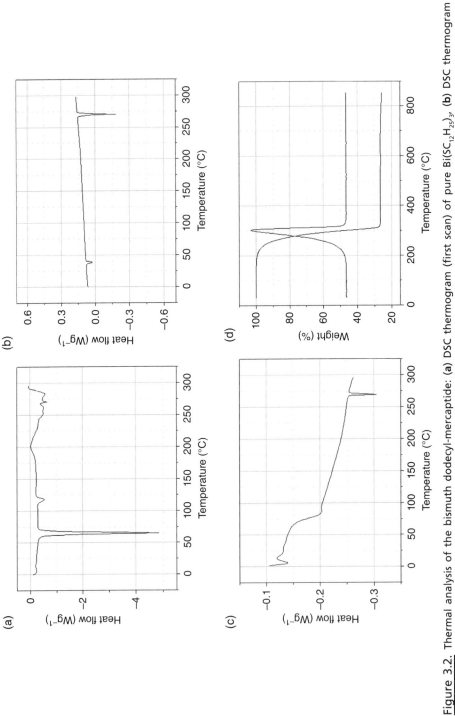

Figure 3.2. Thermal analysis of the bismuth dodecyl-mercaptide: (**a**) DSC thermogram (first scan) of pure $Bi(SC_{12}H_{25})_3$, (**b**) DSC thermogram (second scan) of pure $Bi(SC_{12}H_{25})_3$, (**c**) DSC thermogram of a thermally annealed $Bi(SC_{12}H_{25})_3$/polystyrene blend, and (**d**) TGA thermogram of pure $Bi(SC_{12}H_{25})_3$.

Cobalt Mercaptide. Cobalt acetate ($Co(COOCCH_3)_2$, Aldrich, 99,9%) and dodec-anethiol ($CH_3(CH_2)_{11}$)SH, Aldrich, 99 + %) have been used for the synthesis of cobalt mercaptide. Cobalt acetate was dissolved in ethanol under stirring, and a solution of dodecanethiol in ethanol was added drop by drop to this solution. Then the precipitate was isolated by vacuum filtration and it was washed by ace-tone. Cobalt mercaptide was quite soluble in nonpolar organic solvents, and its thermal decomposition gave a pure CoS phase.

Nickel Mercaptide. Sodium dodecanethiolate ($NaSC_{12}H_{25}$) was prepared by neu-tralization of thiol in ethanol, using a stoichiometric amount of sodium hydroxide. The obtained sodium thiolate was isolated by ethanol evaporation using a Rotavapor apparatus, and then it was dissolved in distilled water and added to an aqueous solution of nickel chloride ($NiCl_2$, Aldrich). A dark-brown waxy solid of $Ni(SC_{12}H_{25})_2$ promptly precipitated and it was isolated by vacuum filtration.

Copper Mercaptide. Copper acetate ($Cu(OOCCH_3)_2$, Aldrich) and dodecanethiol ($CH_3(CH_2)_{11}$SH, Aldrich) were used to synthesize copper(II) dodecyl-mercaptide. The salt was dissolved in ethanol and mixed with an alcoholic solution of thiol at room temperature; a green mercaptide precipitate was isolated by vacuum fil-tration and washed by acetone. Copper(I) dodecyl-mercaptide was obtained in a similar way starting from copper(I) chloride (CuCl, Aldrich).

Many other mercaptides can be obtained using a chemical route similar to that described in the preceding text. However, some mercaptides (e.g., Tl-based mercap-tides) cannot be produced by such simple precipitation method because the metallic ions are promptly reduced by the thiol.

3.4 PRELIMINARY STUDY OF PURE MERCAPTIDE THERMOLYSIS BEHAVIOR BY THERMAL ANALYSIS

Mercaptide thermolysis may behave differently in the presence or absence of polymers [24]; however, in most cases the inorganic phase generated by thermal degradation of mercaptide molecules dissolved in polymer corresponds exactly to that resulting from the thermal degradation of pure mercaptide. Consequently, a preliminary study of pure mercaptide thermolysis by thermal analysis approaches (differential scanning calorim-etry (DSC) and thermogravimetric analysis (TGA)) is usually performed before nano-composite preparation and characterization.

Many thermodynamic information on the behavior of mercaptide thermolysis can be achieved by DSC. In fact, the DSC thermogram of a mercaptide includes details on the number of physical and chemical transformations involved in the thermolysis (i.e., ther-molysis reaction and phase transitions), nature of heats released during the transforma-tions (exo- or endothermic process), and main thermodynamic parameters related to these processes (e.g., temperatures, reaction heats). As an example, Figure 3.2 shows a typical DSC test, consisting in a dynamic run performed on bismuth mercaptide crystalline powder. Differential scanning calorimetry runs were made from room temperature to a temperature higher than the mercaptide decomposition point, under fluxing nitrogen and using sealed aluminum capsules to avoid changes in the thermogram baseline due to the evaporation of organic by-products. In order to identify the phases generated during the mercaptide thermolysis, DSC tests on two references (dodecyl-thioether, m.p. $38\,°C$; dodecyl-disulfide, m.p. $34\,°C$, etc.) were required. The DSC thermogram of pure $Bi(SC_{12}H_{25})_3$ in Figure 3.2 (performed at $10\,°C$ min^{-1}) includes three quite intensive endothermic signals and one broad exothermic signal. The endothermic signal at $64\,°C$ corresponds to the bismuth mercaptide melting point, the less intensive signal at $120\,°C$ is produced by the bismuth mercaptide thermal decomposition, and the very small peaks in the range $269\,°$–$271\,°C$ are probably due to the melting of the produced zero-valence bismuth. The broad exothermic peak appearing in the range $150\,°$–$200\,°C$ is related to the clustering of bismuth atoms coming from the mercaptide molecule thermal degradation. As visible, the generated bismuth phase has a melting point lower than the characteristic value of bulk bismuth (i.e., $271\,°C$); such a phenomenon follows to the metal melting point depression caused by the high surface free-energy content of small metal clusters [14]. If a further DSC run is performed on the same sample (see Fig. 3.2), only two endothermic signals may be detected: the first at $38\,°C$ corresponded to the melting of the organic by-product (thioether/polysulfur mixture) that is produced by the thermal degradation of the disulphide; the second signal was that corresponding to the bismuth melting. When the thermolysis reaction involved mercaptide molecules dissolved in polymer, a similar behavior of DSC thermogram was obtained. Similarly, the second DSC run (see Fig. 3.2c) included the melting of generated bismuth phase in addition to the glass-transition temperature of polystyrene (c. $76.4\,°C$). The absence of further melting signals and the lowering of polystyrene glass-transition temperature is indicative of the organic by-product dissolution in the polymer matrix.

Thermogravimetric analysis of pure mercaptides represents a very simple method to establish the nature of the thermolysis product. In particular, since the organic by-product of thermolysis can be completely removed by evaporation at temperatures close to $300\,°C$ (for dodecyl derivatives), the residual weight of the inorganic solid may correspond to the percentage of metal or sulfide in the mercaptide compound. A simple comparison of the calculated metal percentage in the mercaptide with experimental residual weight allows to establish the type of thermolysis reaction (see Fig. 3.2d). However, such analytical approach may give useful results only by using high-purity (recrystallized) mercaptide samples.

3.5 METAL NANOPARTICLE FORMATION IN POLYMERIC MEDIA

In the preparation of hyperfine solid systems, it is of a fundamental importance to prevent particle aggregation. Owing to the limited Brownian motions in high viscous media, particle aggregation is fully prevented if the nanoscopic solid phase is generated in polymers. Amorphous thermoplastics are characterized by elevated viscosity values also above the glass-transition temperature, and therefore they can be considered as ideal matrices for generating (nucleation/growth) nanoscopic solid phases. This medium should be preferentially selected among those polymers with a glass-transition temperature (T_g) lower than the metal precursor decomposition temperature.

To increase the ability of polymers to protect particles from aggregation, also a certain interaction between polymer and solid surface (polymer physical absorption) is required. Amorphous polystyrene represents a quite good material for cluster formation/growth since glass-transition temperature (c. 80 °C) is closed to usual mercaptide decomposition temperatures (120 °–160 °C) and the polymer has a certain protective ability for the presence of side groups that may be physically absorbed on the electrophilic metal surface (π-electron density donation from phenyl groups to the metal). On the contrary, polyethylene has a low glass-transition temperature value (c. –80 °C) and has no ability to protect particle from aggregation by surface absorption. It has been experimentally proved that mercaptide decomposition in polyethylene medium always leads to a completely aggregate metallic phase.

The type of polymeric medium used for mercaptide decomposition may influence the final shape of the achieved nanoparticles. In particular, the faces of a metal crystal have a different ability to bond nucleophilic species, because the acidity of absorption sites depends on the metal coordination number. The quite low ability of polystyrene side groups to be absorbed on the crystal faces is not enough to discriminate between them. Owing to the ester functions in the poly(vinyl acetate) side groups, these molecules are preferentially absorbed on the most acid faces of the metallic crystals. Polymer absorption creates a diffusion barrier on these crystal faces, thus inhibiting their development and simplifying the polyhedral geometry. To observe such a differential growth of the crystal faces, a significant crystal development is required. For such a reason, mercaptide should be slowly decomposed to generate the metal atoms required to grow nuclei by surface deposition. For example, in the growth of triangular gold plates, the mercaptide thermal decomposition was performed at 160 °C, and the annealing treatment required more than 30 min to allow significant growth of metal crystals with a differential development of crystal faces (see Fig. 3.3). Finally, according to the LaMer model for monodispersed particle formation [25], a single nucleation stage must take place during the process, and then the generated nuclei should grow by addition of gold atoms to the crystal surface. During the growth stage the most acid faces do not significantly develop because of polymer absorption, leading to a simple geometrical shape.

To generate very small metal clusters (i.e., clusters with a size of only a few nanometers), continuous nucleation regimes should be realized during the phase precipitation process. Continuous nucleation requires a high nucleation rate, which can be achieved by decomposing the mercaptide at high temperature (300 °–400 °C). Usually, in addition to the very small size, the generated clusters (nuclei) show a pseudo-spherical shape and result quite monodispersed.

3.6 NANOCOMPOSITE MORPHOLOGY

Because of the very small size of generated nanoparticles (typically from a few nanometers to a few nanometer tens) and their embedded nature, the nanocomposite inner microstructure can be conveniently imaged only by TEM, and in some cases HRTEM is required. Owing to the significant difference in the cross sections of atoms involved in the nanocomposite guest and host phases, usually high-quality TEM micrographs are

(a)

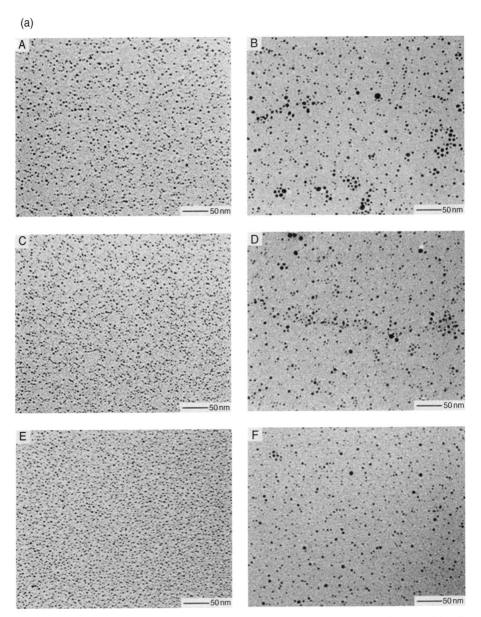

Figure 3.3. (a) Transmission electron microscopy micrographs of $AgSC_{12}H_{25}$/polystyrene blends annealed for 30s (left-side images) and for 180s (right-side images) at 200°C. Different amounts of $AgSC_{12}H_{25}$ were used: 5 wt.% (samples A and B), 10 wt.% (samples C and D), and 15 wt.% (samples E and F). The corresponding particle size distributions are given in the insets.

(b)

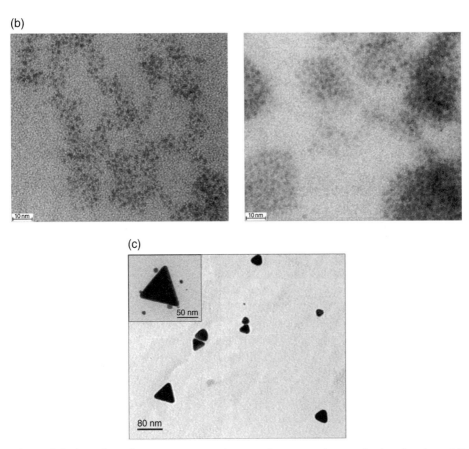

(c)

Figure 3.3. (continued) (b) Transmission electron microscopy micrographs showing the gold/polystyrene nanocomposite inner microstructure (3% by weight of $AuSC_{12}H_{25}$). (c) Transmission electron microscopy micrographs showing the gold/poly(vinyl acetate) nanocomposite inner microstructure (3% by weight of $AuSC_{12}H_{25}$, thermally annealed for 30 min at c. 150 °C).

obtained (highly contrasted images). Transmission electron microscopy investigation gives information about both morphology of embedded particles (e.g., size, shape, orientation) and topology of the particle system (uniform cluster dispersion, presence and type of aggregates, etc.). The TEM-specimen preparation is quite simple: the nanocomposite is dissolved in an adequate organic solvent under sonication, then a drop of this colloidal dispersion is placed on the TEM copper grid, and the solvent is removed by evaporation. Such polymeric layer should be very thin (thickness inferior to 80 nm) to avoid artifacts in the TEM images. Usually, a graphitization stage follows in order to reduce phase contrast and to increase the stability of sample under electron beam. However, for topological investigations, TEM specimens must be prepared by slicing the nanocomposite piece with a cryo-ultramicrotome, since the special topology generated during the annealing treatment can be significantly modified if samples are prepared by dissolution in solvent.

Figure 3.3a shows the internal structure of nanocomposite samples obtained by annealing $AgSC_{12}H_{25}$/polystyrene blends. As visible, a quite uniform, contact-free distribution of metal particles resulted inside these films. Particles have a pseudo-spherical shape with a size of a few nanometers; they are quite monodispersed and without aggregates. However, as shown in Figure 3.3b, for nanocomposites based on gold clusters, when the solvent was quickly evaporated during the blend preparation stage, an unusual morphology characterized by cluster aggregates embedded in polystyrene appeared. In these nanocomposite films, metal clusters were organized in two-dimensional superstructures of different extension. Probably, such special organization of gold clusters was a result of local gradients in the mercaptide concentration. Aggregates were probably generated in the areas where a higher mercaptide concentration was present. Such aggregate topology was observed also in other types of nanocomposite materials. Figure 3.3c shows gold nano-plates developed during the thermal annealing of a $AuSC_{12}H_{25}$/poly(vinyl acetate) blend for 30 min. As visible, nanoparticles of regular shapes are generated in this system, probably for the ability of the acetate groups present in the embedding polymer to selectively bond the different crystallographic faces of growing gold crystals.

3.7 NANOCOMPOSITE STRUCTURAL CHARACTERIZATION

Usually, the inorganic phases generated by thermal decomposition of mercaptide molecules have a crystalline nature, and therefore they can be simply identified by large-angle XRD. Owing to the small size of the crystalline domains (inferior to 50 nm), the diffraction pattern of such polymer-embedded nanocrystals is made of quite broad signals. In addition, the percentage of inorganic phase in a nanocomposite sample is usually low, and consequently peaks are of low intensity. Only those signals corresponding to the most abundant crystallographic planes can be detected, and a low signal/noise ratio characterizes most nanocomposite diffractograms (see Fig. 3.4). Diffraction data of good quality can be achieved only by a slow angular movement of detector and using X-ray sources with adequate anticathode materials. Generally, nanoparticles are single crystals; consequently the broadening of the diffraction peaks allows an approximate evaluation of crystallite size by the Scherrer formula, and the distribution of peak intensities may give also an idea of nanoparticle shape.

The electronic absorption spectra (UV–Vis–near infrared (NIR)) of polymer-embedded metal clusters contain valuable information on their electronic structure and bonding. They illustrate the way in which metallic properties develop in clusters of large-enough size. Unfortunately, there has been little systematic study of the electronic spectra of metal cluster compounds, although this approach can give important structural information. A short description of characterization by UV–Vis spectroscopy of clusters of different nuclearity follows.

Low-nuclearity clusters (i.e., molecular clusters) embedded in polymer show relatively simple electronic absorption spectra that are typical of molecules with well-spaced electronic energy levels. Absorptions of polymer side groups occur in the UV region, and they can be compared with the absorptions of free groups, to study the effect of coordination to metal cluster surface. Metal cluster absorptions occur mainly in the UV and Vis regions, extending in some cases into the NIR. In principle, each absorption

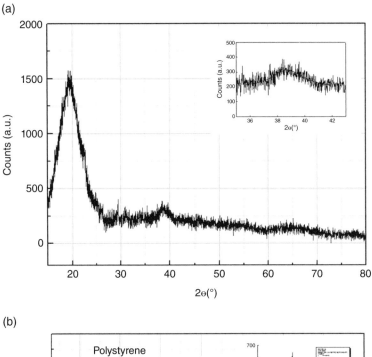

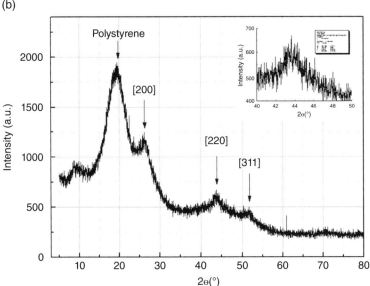

Figure 3.4. Large-angle XRD of (a) gold/polystyrene nanocomposites and (b) CdS/polystyrene nanocomposites.

can be assigned (as $\sigma \rightarrow \sigma^*$) with the longest wavelength absorption corresponding to the HOMO–LUMO gap if this transition is symmetry-allowed. In practice, it can be achieved only for spectra of relatively small clusters. As the clusters get larger, the frontier orbital separation becomes smaller, and the absorption bands move across the Vis region towards the NIR. The quantitative intensities of the longest wavelength absorptions of clusters with nuclearity close to 10 are already very low. Frequently, in addition to one-electron absorption bands in the UV–Vis spectra assigned to transitions between metal–metal bonding and antibonding orbitals, there are bands corresponding to transitions between metal–metal bonding orbitals and ligand antibonding orbitals. On cooling, one-electron absorption bands sharpen because of the effect of the population of vibrational energy levels. λ_{max} may also change, according to the Franck–Condon principle, if the metal–metal distances are not the same in the ground and excited states.

In larger clusters, the resolution of one-electron bands is lost, and a broad, continuous electronic absorption evolves, spanning the Vis region into the NIR. This absorption reflects the overall density of states of the electronic energy levels in the cluster, with the HOMO now being equivalent to the Fermi energy. In gold clusters, it is a $5d \rightarrow 6s, 6p$ transition. At the low-energy end of the spectrum, we can now use the onset wavelength λ_{onset} to estimate the "bandgap" energy difference between the HOMO and LUMO in the cluster. This is a usual practice in the study of small clusters of semiconductors such as CdS, but it has been rarely applied to the electronic spectra of metal clusters. This highlights an unsolved problem in understanding the quantitative density of states at the Fermi level in clusters and colloids, because in colloidal gold the $5d \rightarrow 6s, 6p$ interband transition has comparable intensity only to 1050 nm. At the high-energy end of the spectrum, the shape of the interband absorption in clusters is often obscured by UV absorptions within the polymer side groups. Changing the temperature has little effect on absorption bands of this type. There may be some change in λ_{onset} as the thermal population of levels around the Fermi energy alters.

Delocalized, mobile conduction electrons within a metal particle have a characteristic collective oscillation frequency. This surface plasmon resonance is seen as an absorption band in the UV–Vis spectrum of polymer-embedded clusters. Metals with s-band electronic structures show well-defined plasmon resonances in the Vis region; for silver and gold, these occur respectively at wavelengths of 390 nm and 520 nm. These absorptions weaken as the particle size is reduced, but they can be observed for silver and gold colloids of diameter as small as 10–20 Å. The weak absorption at 510 nm in the Au_{55} spectrum can be a surface plasmon resonance. This 55-atom cuboctahedral gold cluster has an overall diameter (from vertex to opposite vertex) of about 14 Å. The mean first-nearest-neighbor coordination number of the metal atoms, N_1, which is a parameter correlating well with measures of metallic behavior, is 7.85. A similar suggestion had been made for $Ag_{20}Au_{18}$, which shows a strong absorption at 495 nm, the same wavelength as the plasmon absorption in bimetallic AgAu colloids of similar composition. However, the assignment of this band as a plasmon resonance seems unlikely since it is strange that a surface plasmon absorption could be so strong in $Ag_{20}Au_{18}$ but so weak in Au_{55}. The $Ag_{20}Au_{18}$ cluster contains fewer metal atoms than the Au_{55} one, and it is not a close-packed cluster, having a structure based on three 13-atom icosahedra aggregated into an oblate spheroid; its N_1 value is 7.18. Moreover, the effect of particle shape on the plasmon resonance frequency had not been taken into account in comparing the $Ag_{20}Au_{18}$ molecular cluster with bimetallic colloids. In an oblate cluster, any plasmon resonance absorption should occur at a wavelength significantly longer than that in a spherical particle of the same composition, and another effect of the departure from spherical symmetry is to split the resonance into two nondegenerate absorption bands. However, the 510 nm absorption in Au_{55} probably arises from aggregates of cluster molecules in solution, and any plasmon absorption in individual Au_{55} molecules is too weak to be observed above the interband absorption. Au_{55} behaves in an almost identical manner. This limits the number and delocalization of the $6s$ electrons in the Au_{55} clusters. The possible reason for the absence of a clear plasmon absorption in Au_{55} is that the plasmon is broadened and dampened by interaction with the interband absorption, which acts as a decay channel. A further explanation is that the quantitative intensity, C_{abs}, of a plasmon absorption depends on the third power of the particle diameter, a, according to:

$$C_{abs} = \left(\frac{8\pi^2 a^3}{\lambda}\right)\text{Im}\left(\frac{\varepsilon-1}{\varepsilon+2}\right) \tag{3.9}$$

where λ is the wavelength in the medium and ε is the complex relative permittivity of the metal relative to that of the surrounding medium. Even if cluster electrons are fully delocalized and able to undergo a dipole resonance, the plasmon absorption in Au_{55} might simply be too weak to detect above the background interband absorption, which is quite intense at 520 nm. So there may be no need to invoke any special broadening or damping mechanisms. In metals having a less free-electron behavior than silver and gold, absorption maxima in the UV–Vis–NIR spectrum cannot be assigned as pure plasmon resonances, because they also have considerable interband character. In palladium clusters, the UV–Vis band corresponds most closely to a surface plasmon absorption occurring in the UV at about 230 nm. However, the optical spectrum of Pd_{561} (cluster diameter $\approx$30 Å, with $N_1 = 10.05$, assuming an ideal 561-atom cuboctahedral structure) is uninformative, because this spectral region is obscured by electronic transitions within the polymer. A plasmon absorption is not usually detected from platinum clusters, but the theoretical position of this absorption band should be at c. 240 nm. The position and intensity of a surface plasmon absorption band are unlikely to depend on temperature. This gives the possibility of distinguishing experimentally between this type of absorption and one-electron absorptions in the same region of the spectrum.

This outline of polymer-embedded cluster characterization is far from being complete. Advances in nuclear magnetic resonance (NMR), SQUID magnetometry, and many other characterization techniques are of considerable importance as well. In fact, the technological success of a nanostructured material strictly depends on the study of physical properties.

3.8 SOME APPLICATIONS OF METAL–POLYMER NANOCOMPOSITES

Metal–polymer nanocomposites can be exploited for a number of technological applications. The functional uses of these materials are mainly related to their unique combination of high transparency in the Vis spectral range with other physical properties (e.g., luminescence, magnetism, surface plasmon resonance, ultrahigh/infralow refractive index, optical nonlinearity).

Perfectly transparent, lightfast color filters and UV absorbers can be obtained by combining metal clusters of coin metals (silver, gold, etc.) with optical polymers (i.e., amorphous polymers with Vis refractive index close to 1.5, like polystyrene, poly(methyl methacrylate), polycarbonate). The high extinction coefficients that characterize the surface plasmon absorption of these metals allow intensive coloration at very low filling factors, and the nanoscopic filler size makes possible the realization of ultrathin color filters [26, 27].

A system of aggregated particles characterized by surface plasmon resonance absorbs differently from a system of isolated particles. For example, aggregated silver clusters are characterized by a brown coloration, while isolated silver particles (contact-free dispersion of silver clusters) look yellow. Consequently, a change in the interparticle distance produced, for example, by the melting of the absorbed thiolate coating may cause a color switching. Nanocomposites based on polymer-embedded silver clusters derivatized by long-chain alkanethiol show a color change from brown to yellow at a temperature close to the thiolate coating melting point [28, 29].

Magnetic solid phases with nanoscopic dimensions do not scatter Vis light; consequently their embedding into optical plastics leads to magneto-optical materials. Magnetic particles may behave superparamagnetically or ferromagnetically, depending on the size. In particular, when the size is larger than the single magnetic domain, nanoparticles result ferromagnetic; otherwise, they are superparamagnetic. As a consequence, superparamagnetic and ferromagnetic optical plastics can be produced [30, 31]. If particles with acicular shape are embedded in polymer, the magnetic properties can be significantly enhanced by a shape anisotropy effect. These materials can be exploited for a number of technological applications based on their transparency and/or low density (magneto-optics, magnetic levitation, optically transparent data storage systems, etc.).

Luminescent plastics are useful for a number of technological applications (e.g., photoelectric sensing, anticounterfeiting, materials for chip-on-board technology, optical filters for converting high-energy solar light to useful radiation for photovoltaic cells, advanced materials for greenhouse windows (agriculture)) [32]. Plastics can be made luminescent simply by embedding small metal and semiconductor clusters (e.g., sulfides). Such luminescence type is very attractive mainly for the possibility to control the bandgap through material composition and size. Nanoparticles are characterized by a size- and composition-dependent bandgap that can be tuned atom by atom during the synthesis stage to emit at any Vis or infrared wavelength. Polymer-embedded nanoclusters are ready to be used with the great advantage of a very versatile manufacturing of polymers.

Further applications of metal–polymer nanocomposites are in the fields of ultrahigh/infralow refractive index materials [33–35], dichroic color filters [36, 37], nonlinear optical filters [38], catalytic polymer membranes [39, 40], etc.

3.9 CONCLUSIONS

Metal nanoparticles embedded in polymeric matrices represent a novel nanostructured material class. These materials combine optical transparency to magnetism, luminescence, UV–Vis absorption, thermochromism, etc., leading to unique functional materials that can be conveniently exploited for a number of applications in different technological fields. A really effective synthesis route for these materials is represented by the thermal decomposition of mercaptide molecules dissolved in polymer. Mercaptides can be dissolved/dispersed in polymers and thermally degraded at temperatures compatible with polymer stability (100°–250°C), generating metal atoms or metal sulfide molecules that lead to very small nanoparticles by clustering. Mercaptide synthesis, blending with polymers, and thermal decomposition are quite simple and general operations; consequently this approach can be easily applied for the preparation of a variety of nanocomposite systems.

REFERENCES

[1] W. Caseri, *Macromol. Rapid Commun.* **21**(2000)705–722.

[2] [2]A.B.R. Mayer, *Mater. Sci. Eng. C* **6**(1998)155–166.

[3] G. Carotenuto et al., EP 1489133A1 (2004).

[4] G. Carotenuto et al., U.S. Patent, 2006/0121262A1 (2006).

[5] Y. Hasegawa et al., U.S. Patent, 5,110,622 (1992).

[6] A. Davlin, U.S. Patent, 6,231,925 (2001).

[7] B. Howard et al., U.S. Patent, 2,994,614 (1961).

[8] K.-D. Fritsche et al., U.S. Patent, 5,707,436 (1998).

[9] B. Howard et al., U.S. Patent, 2,984,575 (1961).

[10] H. Kermit et al., U.S. Patent, 2,490,399 (1949).

[11] P.H. Nguyen et al., U.S. Patent, 4,808,274 (1989).

[12] G. Carotenuto, L. Nicolais, *J. Mater. Chem.* **13**(2003)1038–1041.

[13] G. Carotenuto, L. Nicolais, P. Perlo, *Polym. Eng. Sci.* **46(8)**(2006)1016–1020.

[14] G. Carotenuto, L. Nicolais, "Nanocomposites, Metal-Filled," in Encyclopedia of Polymer Science and Technology, 3rd Edition, John Wiley & Sons, Inc. Hoboken, NJ, USA, 2003.

[15] T.H. Larsen, M. Sigman, A. Ghezelbash, R.C. Doty, B.A. Korgel, *J. Am. Chem. Soc.* **125**(2003)5638–5639.

[16] M.B. Sigman, A. Ghezelbash, T. Hanrath, A.E. Saunders, F. Lee, B. Korgel, *J. Am. Chem. Soc.* **125**(2003)16050–16057.

[17] A. Ghezelbash, M.B. Sigman, B.A. Korgel, *Nano Lett.* **4(4)**(2004)537–542.

[18] A.P.G. de Sousa, R.M. Silva, A. Cesar, J.L. Wardell, J.C. Huffman, A. Abras, *J. Organomet. Chem.* **605**(2000)82–88.

[19] L. Qu, W. Tian, W. Shu, *Polym. Degrad. Stab.* **76**(2002)185–189.

[20] K.M. Anderson, C.J. Baylies, A.H.M. Jahan, N.C. Norman, A.G. Orpen, J. Starbuck, *Dalton Trans.* 2003, 3270–3277.

[21] W. Clegg, M.R.J. Elsegood, L.J. Farrugia, F.J. Lawlor, N.C. Norman, A.J. Scott, *J. Chem. Soc. Dalton Trans.* 1995, 2129–2135.

[22] V.K. LaMer, R.H. Dinegar, *J. Am. Chem. Soc.* **72(11)**(1950)4847–4854.

[23] G. Carotenuto, G. LaPeruta, L.Nicolais, *Sens. Actuators B Chem.* **114**(2006)1092–1095.

[24] P. Conte, G. Carotenuto, A. Piccolo, P. Perlo, L. Nicolais, *J. Mater. Chem.* **17**(2007)201–205.

[25] I.G. Dance, K.J. Fisher, R.M. Herath Banda, M.L. Scudder, *Inorg. Chem.* **30**(1991) 183–187.

[26] M. Zheng, M. Gu, Y. Jin, G. Jin, *Mater. Res. Bull.* **36**(2001)853–859.

[27] G. Carotenuto, *Appl. Organomet. Chem.* **15**(2001)344–351.

[28] G. Carotenuto, G. LaPeruta, L. Nicolais, *Sens. Actuators B Chem.* **114**(2006)1092–1095.

[29] G. Carotenuto, F. Nicolais, *Materials* **2**(2009)1323–1340.

[30] G. Carotenuto, G. Pepe, D. Davino, B. Martorana, P. Perlo, D. Acierno, L. Nicolais, *Microw. Opt. Technol. Lett.* **48(12)**(2006)2505–2508.

[31] K.E. Gonsalves, G. Carlson, M. Benaissa, M. Jose-Yacaman, D.Y. Kim, J. Kumar, *J. Mater. Chem.* **7(5)**(1997)703–704.

[32] G. Carotenuto, A. Longo, P. Repetto, P. Perlo, L. Ambrosio, *Sens. Actuators. B Chem.* **125**(2007)202–206.

[33] M. Weibel, W. Caseri, U.W. Suter, H. Kiess, E. Wehrli, *Polym. Adv. Technol.* **2**(1991)75–80.

[34] L. Zimmerman, M. Weibel, W. Caseri, U.W. Suter, P. Walther, *Polym. Adv. Technol.* **4**(1992)1–7.

[35] L. Zimmerman, M. Weibel, W. Caseri, U.W. Suter, *J. Mater. Res.* **8**(1993)1742–1748.

[36] Y. Dirix, C. Bastiaansen, W. Caseri, P. Smith, *J. Mater. Sci.* **34**(1999)3859–3866.

[37] Y. Dirix, C. Bastiaansen, W. Caseri, P. Smith, *Adv. Mater.* **11**(1999)223–227.

[38] S. Qu, Y. Song, H. Liu, Y. Wang, Y. Gao, S. Liu, X. Zhang, Y. Li, D. Zhu, *Opt. Comm.* **203**(2002)283–288.

[39] L. Troger, H. Hunnefeld, S. Nunes, M. Oehring, D. Fritsch, *Z. Phys. D* **40**(1997)81–83.

[40] D. Fritsch, K.V. Peinemann, *Catal. Today* **25**(1995)277–283.

4

MACROMOLECULAR METAL CARBOXYLATES AS PRECURSORS OF METALLOPOLYMER NANOCOMPOSITES

G. I. Dzhardimalieva[1] and A. D. Pomogailo[2]

[1] *Institute of Problems of Chemical Physics, Russian Academy of Sciences, Chernogolovka, Moscow, Russia*
[2] *Laboratory of Metallopolymers, Institute of Problems of Chemical Physics, Russian Academy of Sciences, Chernogolovka, Moscow, Russia*

4.1 INTRODUCTION

Metal carboxylates are widely used in science and technology. They are a part of polynuclear coordination compounds (in catalytic and biomimetic systems) and metal proteins, intermediate compounds of many metabolic processes. Biochemical behavior of metal enzymes and antibodies is determined in many respects by their carboxylate function [1, 2]. Among numerous metal-carboxylate derivatives, a special place is occupied by salts of unsaturated carboxylic acids: acrylic, methacrylic, crotonic, oleic, fumaric, maleic, acetylenedicarboxylic, vinylbenzoic acids, and so on, which are typical metal-containing monomers because they have a multiple bond able to be cleaved and a metal ion chemically bound to the organic part of the molecule [3]. In recent years polymerization of such monomers has been the subject of intense studies due to the practical importance of the resulting products and their compositions. The properties of polymers

Nanocomposites: In Situ *Synthesis of Polymer-Embedded Nanostructures*, First Edition.
Edited by Luigi Nicolais and Gianfranco Carotenuto.
© 2014 John Wiley & Sons, Inc. Published 2014 by John Wiley & Sons, Inc.

in which every repeating unit contains an equivalent of the metal differ appreciably from the properties of traditional polymeric materials. The presence of reactive functional groups and the diversity of the coordination environment of the metal create unique possibilities for the design of novel advanced materials. Fairly promising in this respect are metal oxoclusters containing unsaturated carboxylate groups, which can serve as nanostructure elements for the production of organic–inorganic hybrid nanocomposites [4]. These are highly organized systems with strictly defined sizes and shapes, which are retained in the final products, and therefore it is possible to attain their homogeneous distribution in the material and obtain monodisperse nanostructures.

There are currently three basic approaches to the preparation of metal-containing polymers [5]: (i) reactions of metal compounds with functionalized linear polymers in which the main polymer chain remains unchanged (so-called polymer-analogous conversions); (ii) polycondensation of appropriate precursors, a process in which a metal ion is incorporated into the main chain and its removal leads to destruction of the polymer; and (iii) polymerization and copolymerization of metal-containing monomers. Homo- and copolymerization of unsaturated metal carboxylates are the important methods for obtaining of macromolecular metal carboxylates. In most cases, polymeric metal salts are prepared by radical polymerization, which comprises the same elementary steps as for conventional monomers. The radical polymerization of the salts of unsaturated carboxylic acids can be induced by any initiator or initiating radiation. However, azobis(isobutyronitrile), benzoyl peroxide, potassium or ammonium persulfates, H_2O_2, tert-butylhydroperoxide, and various redox systems are used most often. Polymerization of metal carboxylates in the solid state [6–9] and under matrix devitrification conditions [10] is often initiated by γ-radiation. Chelated complexes of alkylcobalt with tridentate Schiff bases appeared to be efficient initiators of the low-temperature radical polymerization of some metal acrylates [11]. Examples of photoinduced polymerization are known as well. Tetraethoxytitanium(IV) methacrylate derivatives are successfully polymerized in thin layers or on the surface of a metallic substrate upon UV irradiation [12]. The example of atom transfer radical polymerization (ATRP) for the metal-containing monomers has been reported for sodium methacrylate. The process occurs in an aqueous solution at 363 K in the presence of a macroinitiator based on poly(ethylene oxide), copper(I) bromide catalyst, and 2,2-bipyridine in a 2:2:5 molar ratio [13]. At pH < 6, the reaction was ineffective, which may be due to protonation of bipyridine and the lack of solubility of the catalyst under these conditions. The resulting poly(ethylene oxide-block-sodium methacrylate) copolymer had a relatively low molecular mass and narrow polydispersity (1.2–1.3). Recently a similar method has been used to carry out polymerization of sodium methacrylate on the surface of various substrates modified by an initiator of ATRP [14, 15]. This procedure gives polyelectrolyte layers of controlled composition, thickness, and density.

As concerns polymer-analogous conversions, all organic polyacids are capable to form in alkaline, neutral, and low acidic media quite strong metal complexes. Thus, polyacrylic acid (PAA) binds Cu(II) much stronger than its low molecular weight analogs. This process can proceed with ionized as well as with nonionized carboxylic groups. In the former case, contribution from ionic component into the total coordination bond energy becomes the definitive [16, 17]. Carboxyl group-containing polymers can be utilized in complexation processes in combining with micro- and ultrafiltration [18]. For example, PAA (mol. mass 30,000) is used for the removal of Zn^{2+} and Ni^{2+} ions from water on a polysulfonic membrane with retention coefficient of 97–99% [19]. This process is competitive with osmosis, nanofiltration, electrodialysis, liquid membranes, and so on.

Binding of metal with complexing agents of natural systems is of particular importance, since functioning of transitional metal ions in biological systems, transport, and assimilation of metal ions in living organisms are based on the binding of the ions by functional groups of biopolymers. Polycarboxylates are widely used in interdisciplinary areas, such as ecotoxicology, water chemistry, and plant nutrition; they are included into protein formations. Thus, interaction of Ca^{2+} ions with COO^- groups of a protein results in a change of biopolymer swelling and is a critical stage of blood coagulation, irritation and contraction of nerves and muscles, and cell movement [20, 21]. These macroligands have received broad utilization for the preparation of new types of detoxicants, immobilized ferments, pharmaceuticals, and so on.

4.2 STRUCTURE AND MOLECULAR ORGANIZATION OF MACROMOLECULAR METAL CARBOXYLATES

The properties of (co)polymer-based materials are known to be substantially dependent on the sequence of different units in the polymer chains. The differences in the distribution of units are reflected in the type of intermolecular interactions; they are responsible for the diversity of supramolecular structures and affect the properties of materials.

4.2.1 Metal-Carboxylate Ionomers

Many properties of metallopolymers are determined by ion aggregation. This is especially pronounced for alkali and alkaline earth metals, which allows them to be considered as ionomers.

The formation of stable ion pair aggregates [22], which may form either multiplets or clusters depending on the concentration, is believed to be the key factor determining the structure and properties of ionomers. Multiplets are small compact aggregates of ion pairs, while clusters are aggregates formed from a number of multiplets. It is assumed [23] that a region with lower mobility of polymeric chains appears around each multiplet. In the case of high concentrations of charged groups in the polymeric matrix (>5–7%), these regions start to overlap, thus forming extended ion-enriched structures called domains (or clusters), which often behave as an individual phase. The contents of separate aggregates are determined by physicochemical methods and estimated from dynamic, mechanical, rheological, or dielectric properties of metallopolymers.

In the Raman spectra of sodium polymethacrylate [24], the absorption band at $254\,cm^{-1}$ was assigned to CO vibrations in multiplets, and the band at $166\,cm^{-1}$ was assigned to their vibrations in clusters. This is due to the fact that in complex aggregates the electrostatic interactions are more shielded and the frequencies of ion vibrations decrease with respect to the principal vibrational band for a particular cation or multiplet.

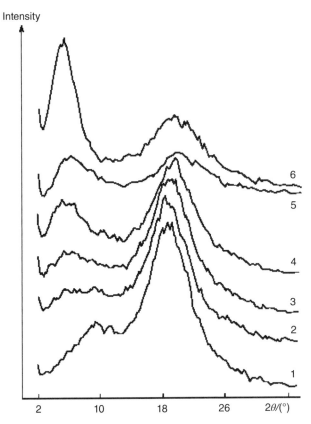

Figure 4.1. Wide-angle X-ray scattering spectra of zinc acrylate–styrene copolymer with a salt content of 3.67 (1), 5.51 (2), 7.02 (3), 9.83 (4), 17.59 (5), and 19.48 mol% (6). Reproduced from Reference [27] by permission of the American Chemical Society.

Analogous regularities were also observed in the case of alkali metal acrylate copolymer with styrene [25]: the vibrational bands of the COO^- group in the clusters occurred at about 155 cm^{-1} for K$^+$ and Na$^+$ ions and at 95 cm^{-1} for the Cs$^+$ ions. These aggregates were described in detail [26]. It was shown by wide-angle X-ray scattering (WAXS) and SAXS [27] that in the zinc acrylate–styrene copolymer, small ionic aggregates are mainly formed. In the WAXS spectrum, the ion peak at $2\Theta = 5.58$ appears at a zinc acrylate concentration of 7.02 mol%, and its intensity increases with an increase in the salt content (Fig. 4.1). Alkali metal acrylates copolymerized with styrene form 70–100 Å cluster aggregates in addition to ion pairs and multiplets [28, 29] even at concentrations of 3.85 and 5.16 mol%. This is also typical of other Zn-containing ionomers [26, 30]. For example, in ethylene–methacrylic acid copolymers (the degree of neutralization with zinc of 0.32–0.83), the size of ionic aggregates equals 0.45 nm over the whole range of metal ion concentrations; the same is true for the copper and iron ethylene ionomers.

Character of the ionic associations is also determined by the nature of a surrounding polymeric matrix. Removal of the charged groups from the main chain can favor the formation of ionic multiplets owing to the reduction of steric hindrances. Poly(ethyl acrylate-co-itaconate) containing two ionic groups in the one unit

$$
\begin{array}{c}
\text{COO--Na}^+ \\
| \\
\left[\text{CH}_2 - \underset{\underset{\text{COOC}_2\text{H}_5}{|}}{\text{CH}} \right]_x \left[\text{CH}_2 - \underset{\underset{\underset{\underset{\text{COO--Na}^+}{|}}{\text{CH}_2}}{|}}{\text{C}} \right]_y
\end{array}
\qquad (4.1)
$$

shows a high degree of clusters formation in comparison with a similar ionomer based on the polystyrene characterized by the formation of several multiplets only. It is confirmed by the values of the relaxation module and the tangent of the angle of mechanical and dielectric losses of the copolymers given [31]. Such behavior is connected with the noticeably low value of glass-transition temperature T_g of the polyethylacrylate matrix against the polystyrene system (~398 K) [32]. According to the dynamic–mechanical thermal analysis data [33], the ionomeric copolymers of polyethylacrylate and acrylic acid (poly(ethylacrylate)-co-acrylic acid (3.6–15.2 mol. %)), neutralized by various cations, are characterized by two glass-transition temperatures. Lower glass points correspond to the T_g of the polyethylacrylate matrix, while temperature transfers revealed in the high-temperature area are caused by cluster aggregates [34–36].

Sizes of the forming domains can vary in a wide range depending on many factors, including neutralization degree of carboxyl groups, coordination of a metal, a level of microphase division, and also nature and molecular weight of an ionomeric molecule. In some cases, such ion-containing polymers can be considered as nanocomposites on a molecular level. For example, according to the transmission electron microscopy (TEM) data for the K-maleate ionomeric three-block copolymer styrene–butadiene–styrene, the diameter of ionic domains is 3–8 nm (Fig. 4.2) [37].

It is well known that in selective solvents, block copolymers of the PS-PAA type (block ionomers) exist as reversed micelles. In organic solvents, block copolymers are segregated into microphases with spherical, cylindrical, and lamellar morphology. Metal ions are bound with carboxyl groups of the micelle core due to formation of covalent or ionic bonds. Recently more complex block copolymers have been developed for these aims, for example, diblock copolymer (methyltetraclododecene)$_{400}$ (2-norbornene-5,6-dicarboxylic acid)$_{50}$ [38]. Ions of Ag, Au, Cu, Ni, Pb, Pd, Pt, and others form macrocomplex with units of the micelle core (Fig. 4.3). The loading of a metal can reach quite significant values, over 1 g/g of PAA block under the optimal conditions (Fig. 4.4).

It is necessary to note that state of the ionic aggregation depends on many factors, for example, on the cation nature, on the polymeric chain microstructure, and on ways of the ionomer obtaining. For example, metal-containing polymers obtained by solid-phase polymerization under the action of high pressures in combination with shearing

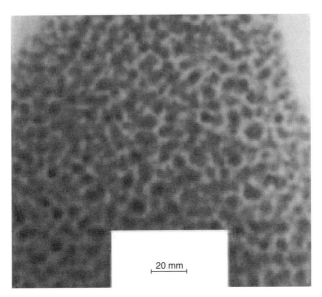

Figure 4.2. Transmission electron microscopy microphotograph of lead-containing maleate ionomers of styrene–butadiene–styrene triblock copolymer. Reproduced from Reference [37] by permission of the American Chemical Society.

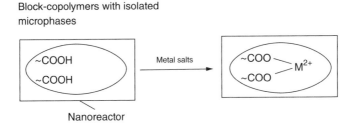

Figure 4.3. A principal scheme of the block copolymer and its metallocomplex formation. Reproduced from Reference [38] by permission of the American Chemical Society.

deformations (HP+SD) show strong antiferromagnetic exchange between paramagnetic centers, both in homopolymers and in heterometallic copolymers on the basis of Ni(II), Cu(II), and Ti(IV) acrylates (Table 4.1) [39, 40]. Antiferromagnetic exchange is connected, most probably, with the interchain interactions of the paramagnetic centers developed as a result of conformational changes in macrochains under the action of HP+SD; that is, the favorable conditions for the formation of multiplet and cluster domain structures are created. It is typical that polymers and copolymers obtained by liquid-phase polymerization do not show antiferromagnetic exchange after HP+SD treatment. It confirms that structure of the complexes, combined into clusters with anti-ferromagnetic interaction, is formed at the stage of copolymer formation in afterflow conditions.

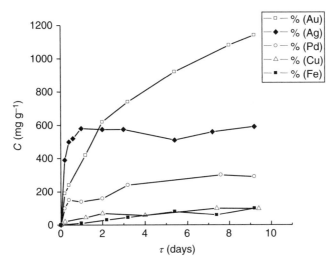

Figure 4.4. The change of the degree of metal ion loading by block copolymer of (methyltetracyclododecene)$_{400}$(2-norbornene-5,6-dicarboxylic acid)$_{50}$ versus time (C is the loading capacity, mg of metalg^{-1} of polymer). Reproduced from Reference [38] by permission of the American Chemical Society.

TABLE 4.1. Magnetic Properties of Heterometal-Containing Copolymers

Copolymer	Method of synthesis	The content of M_1, mol.%	μ_{ef}, B.M.		Antiferromagnetic exchange
			295 K	80 K	
Ni(II) acrylate (M_1)–Cp$_2$Ti(MAA)$_2$	Radical, in solution	54	3.40	3.36	No exchange
The same	The same	84	3.38	3.30	The same
''	''	92	3.27	3.18	''
Homopolymer of Ni(II) acrylate	''	100	3.29	3.28	''
Ni(II) acrylate (M_1)–Cp$_2$Ti(MAA)$_2$	Solid phase (HP+SD)	42	4.30	3.75	Exchange
The same	The same	76	4.05	3.38	The same
''	''	91	3.73	3.36	''
Homopolymer of Ni(II) acrylate	''	100	4.73	3.78	''
Cu(II) acrylate (M_1)–Cp$_2$Ti(MAA)$_2$	''	39	1.58	1.05	Strong exchange
The same	The same	62	1.57	1.03	The same
''	''	73	2.53	1.56	''
''	''	84	1.48	1.07	''
Homopolymer of Cu(II) acrylate	''	100	1.42	1.15	''

4.2.2 Hybrid Supramolecular Structures

The binuclear metal carboxylates $M_2(O_2CR)_4L_2$ tend to form lantern-type structures. Polymerization of binuclear carboxylates with linear dicarboxylate bridges yields highly symmetric 2D structures, and the interplane contacts in such systems produce an infinite chain of linear micropores (Fig. 4.5). This structure is characteristic of both fumarates and trans,trans-muconates containing binuclear Mo_2^{4+} [41–43], Rh_2^{4+} [44], and $Ru_2^{2+,3+}$ [45] units. The synthesis of this type of dimetal carboxylates in the presence of linear polymers, for example, poly(ethylene glycol), gives rise to supramolecular inclusion complexes [41, 42].

Among promising trends in the synthesis of polymeric supramolecular structures, note intercalation of acrylate ions followed by their polymerization *in situ* in molecules of layered double metal hydroxides (LDH) as inorganic "hosts" [46, 47]. Using this approach, acrylate and polyacrylate nanocomposites containing $Ni_{0.7}M_{0.3}$ carboxylate fragments (M = Fe, Co, Mn) were obtained [48, 49]. In the case of Fe intercalates, it is possible to isolate monomeric (and polymeric) systems in successive synthetic stages, whereas for Co- and Mn-containing nanocomposites, intercalation and polymerization take place in one stage to give polyacrylate structures. Depending on the conditions of synthesis, the interlayer distances in these materials are 7.8–12.5Å.

In addition to micro- and mesoporous compounds, liquid and organic crystals, micelles, bilayer lipids, and so on are often used as highly organized media for the preparation of supramolecular systems [50]. Thus, mixing of poly(p-phenylenevinylene) (PPV) with amphiphilic acrylate monomers or their precursors, for example, salt [51, 52],

$$Na^+ \; {}^-O_2C \; \overset{\displaystyle O(CH_2)_{11}O_2CCH=CH_2}{\underset{\displaystyle O(CH_2)_{11}O_2CCH=CH_2}{\bigcirc \; O(CH_2)_{11}O_2CCH=CH_2}}$$

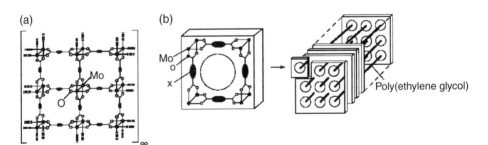

Figure 4.5. Structure of coordination polymers based on (a) dimetal carboxylates and (b) scheme of formation of linear micropores, which can accommodate poly(ethylene glycol) chains. Reproduced from Reference [41] by permission of the American Chemical Society.

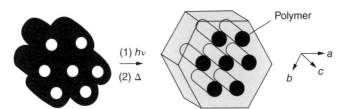

Figure 4.6. Scheme of stabilization of lyotropic liquid-crystalline phase by polymerization of an acrylate monomer. Reproduced from Reference [51] by permission of the American Chemical Society.

affords a reverse lyotropic liquid-crystalline phase in which the polymer, PPV, pierces the hexagonal structure and is oriented parallel to the *c* axis (Fig. 4.6).

Photopolymerization of the acrylate salt and subsequent heating of the cross-linked matrix lead to the formation of a hybrid material with more intensive fluorescence compared to the bulk PPV. Moreover, a new intensive band in the emission spectrum at 670 nm appears in case of Eu(III) salt [52]. It testifies interaction of metal cation with PPV chains with possible energy transport between components. Sizes of the inverted hexagonal phase depend on the metal nature and type of the metal-carboxylate interaction [52, 53] and are also determined by length and structure of the aliphatic part of the amphiphilic mesogenic metal-containing monomer [54] *p*-styryl- or *p*-styryloxy-octadecanoates.

A similar approach to the synthesis of highly organized systems was applied to prepare polymeric micelles based on the 4-vinylbenzoate monomer:

$$\text{CH}_3$$
$$\text{—COO}^- \ \text{H}_3\text{C}-\overset{|}{\underset{|}{\text{N}^+}}-(\text{CH}_2)_{11}\text{CH}_3$$
$$\text{CH}_3$$

The rodlike shape of molecule of the surfactant molecule is retained during radical polymerization, and the product is thermally stable and is not changed on dilution [55].

These data imply that the molecular organization of metallopolymers can be different: from linear polymers to two- and three-dimensional networks and supramolecular structures. The structural and chemical control of polymerization at all levels of metallo(co)polymer organization, that is, molecular, topological, and supramolecular levels, allows the production of new materials with a set of valuable properties.

4.3 PREPARATION OF METALLOPOLYMER NANOCOMPOSITES BASED ON METAL CARBOXYLATES

The interest in metallopolymer nanocomposites is due to the unique combination of properties of metal-containing nanoparticles with mechanical, film-forming, and other polymer characteristics, which allows these materials to be used as magnetic devices for data recording and storage, catalysts, and sensors and for various purposes in medicine and biology [56].

Homo- and copolymers of acrylic and methacrylic acids and their salts are often used for stabilization of metal-containing dispersions. For example, PbS nanocomposites with a styrene–methacrylic acid copolymer [57] and with an ethylene–methacrylic acid copolymer [58] and CuS nanocomposite with a poly(vinyl alcohol)–polyacrylic acid copolymer [59] were prepared. Such examples are very numerous. On the one hand, carboxylated compounds of a monomeric and polymeric structure can be molecular precursors of nanocomposite materials. On the other hand, carboxyl groups of macroligands are efficient stabilizers of nanoparticles; these functions are frequently developed together in one system.

4.3.1 Controlled Thermolysis of Unsaturated Metal Carboxylates

One of the perspective obtaining methods of metal-containing nanoparticles and their polymeric composites is thermal conversions of metal-containing monomers. It is possible to combine *in situ* formation of superfine metal particles and stabilize the polymeric matrix during these thermal conversions [60–63]. Metal-containing polymeric nanocomposites on the basis of metal acrylates [64–67], their cocrystallizates [68], and also maleates [69, 70] were obtained using such approach. Microstructure of the formed composites is represented by the metal-containing nanoparticles of 5–30 nm diameter, and close to spherical form, they are dispersed homogeneously in the polymeric matrix with average distance of 10–12 nm [71] (Fig. 4.7). Uniformity of distribution of the metal-containing particles in the matrix and their narrow size distribution testifies, apparently, a big degree of homogeneity of processes of decarboxylation and formation of a new phase. It is important to note that the average size of the particles, forming

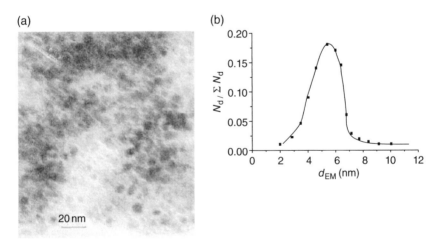

Figure 4.7. Transmission electron microscopy (a) microphotograph (b) and diagram of nanoparticle distribution on the size for the products of thermolysis of the Fe(III)–Co(II) acrylic cocrystallizate at 643 K. Reproduced by permission from Reference [40].

during thermal conversions of unsaturated metal carboxylates, is lower than for the products of thermal conversions of saturated metal carboxylates [72].

Systematic research of thermolysis of unsaturated metal carboxylates allowed to reveal a community of character of their conversions, consisting in sequences of three basic macrostages [73]:

1. Dehydration of crystalline hydrates of monomers ($T_{term} < 423$ K) with simultaneous reorganization of ligand environment, accompanied by separation of a part of carboxylated ligands.
2. Solid-phase polymerization of the reorganized dehydrated monomer ($T_{term} \approx$ 453–493 K).
3. Decarboxylation of the formed (co)polymer at high temperatures ($T_{term} > 473$ K). Main gas evolution and mass loss of a sample at thermolysis are connected with the last process.

Oleates and octanoates of metals are the most frequently used molecular precursors of nanostructured materials among other monomeric carboxylates. Thermolysis of these complexes in combination with surfactants and other reagents is usually carried out in a solution of high-boiling solvents (octadecane, octadecene, docosane, octyl ether, etc.). Doubtless advantages of thermal decomposition of carboxylated compounds in an inert solvent are the opportunity of the controlled synthesis of practically monodisperse nanocrystals with a high yield, narrow size distribution, and high crystallinity [74–78]. Nanocrystals of semiconductor metal sulfides were obtained by thermal decomposition of metal–oleate complexes in an alkanethiol [79, 80].

4.3.2 Metal Nanoparticles in Polymer Carboxylate Gels and Block Copolymers

Amphiphilic diblock copolymers, for example, polystyrene-block-polyacrylic acid, in organic and aqueous solutions are widely used for the encapsulation of semiconductor nanoparticles of metal sulfides at the stage of their formation [81, 82]. One of such examples is interesting by the fact that di- and three-block copolymers give unique opportunities for fine regulation not only of sizes of nanoparticles but also regulation of morphology of metal-containing polymeric nanocomposites on their basis. Quantum dots of CdS were obtained in the micelles of the three-block copolymer of poly(ethylene oxide)-block-polystyrene-block-poly(acrylic acid) of various architecture [83]. The advantage of many polymeric gels and nanocomposites on their basis is their biocompatibility; due to this fact, they can be used in medicine for creation of carriers of medicinal substances and for their transportation. The network structure of microgels providing probability of nucleation and growth of nanoparticles in each void and high sensitivity of these systems to changes of external factors also has important value. Various types of polymeric microgel nanocomposites containing metallic, magnetic, semiconductor, ceramic, and other nanoparticles have been developed at present time [84, 85]. For example, hydrodynamical radius of particles of the microgel of poly (N-isopropylacrylamide-acrylic acid-2-hydroxyethylacrylate) [86] in the region of

$2.3 < pH < 9.2$ increases from 230 to 600 nm that results in an increase in the CdS content from 0.04 to 0.12 g/g. Nanocomposite microspheres of ZnS- and CdS-poly (*N*-polyisopropylacrylamide-*co*-methacrylic acid) reveal very interesting superficial morphology in the form of figured structures that is connected with nonuniform precipitation of metal sulfide because of unhomogeneous distribution of metal ions within the microgel [87].

4.3.3 Sol–Gel Methods in the Obtaining of Oxocluster Hybrid Materials

In recent years, considerable attention has been paid to material based on unsaturated metal oxo-carboxylates. Owing to the relatively small size (~1 nm), oxoclusters can be used as structural elements of hybrid organic–inorganic nanocomposites [88, 89]. As the initial component, one can use oxo-carboxylate clusters containing functional groups that accomplish covalent bonding between the metal oxocluster fragment and the polymer chain. In this case, the inorganic core is connected to the unsaturated fragments of the hybrid molecule through electrostatic interactions and hydrogen bonds, as this is shown for the oxocluster $[(Bu^nSn)_{12}O_{14}(OH)_6](O_2CC(Me)=CH_2)_2$ [90]. Its copolymerization with MMA gives the polymethyl methacrylate–methacrylate copolymer crosslinked by the $[(Bu^nSn)_{12}O_{14}(OH)_6]$ units, which does not change during polymerization transformations.

For the preparation of hybrid thin films based on silica gel with inserted hafnium oxoclusters, the oxocluster $Hf_4O_2(O_2CC(Me)=CH_2)_{12}$ methacrylate and 3-methacryloyloxypropyltrimethoxysilane were used [91, 92]. The chemical binding of the components was performed by photochemical polymerization of methacrylate groups; the silane alkoxy groups were pre-hydrolyzed and pre-condensed to give the oxide network.

4.4 METAL-CARBOXYLATE NANOCOMPOSITE MATERIALS

Metal-containing polymeric nanocomposite materials have interesting magnetic, catalytic, optical, and other properties in dependence on the nature of a metal-containing dispersed phase.

In particular, owing to the formation of ferrimagnetic Fe_3O_4 and $CoFe_2O_4$ nanoparticles, the thermolysis products of metal acrylates and cocrystallizates of cobalt(II) and iron(III) acrylates behave as solid magnets with a coercive force of 0.18 T and residual magnetization of 15.5 mT at room temperature (Fig. 4.8) [65, 71]. A twice higher coercive force was found for Co-containing particles (average diameter of 7 nm) composed mainly of CoO prepared from cobalt(II) acrylate in a polymeric matrix. The spinel ferrites $Zn_{0.5}Mn_{0.5}Fe_2O_4$ [93], $CoFe_2O_4$ [94], and $MnFe_2O_4$ [95] synthesized by pyrolysis of the corresponding polyacrylate salts exhibit superparamagnetic properties at room temperature and can be used as magnetic and radar absorbent materials.

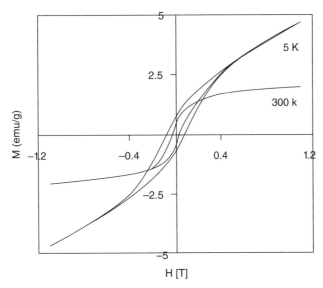

Figure 4.8. ZFC hysteresis loops, recorded at 5 K and 300 K for the product of thermolysis of Co(II) acrylate at 663 K. Reproduced by permission from Reference [40].

Unsaturated metal carboxylates are of considerable interest as precursors in the design of multimetallic superconducting ceramics, multicomponent alloys, oxide electrodes, and so on. Traditional methods for batch preparation of high-temperature superconducting (HTSC) ceramics are poorly reproducible, often accompanied by the formation of micro- and macroinhomogeneities, in particular, nonconducting phases. The synthesis of multicomponent ceramics based on polymers allows one to overcome these drawbacks and obtain structurally homogeneous products. This is attained, for example, by the use of a polymeric matrix with metal ions dispersed there down to the molecular level. This method was used to prepare $YBa_2Cu_3O_{7-x}$ HTSC ceramics from polymeric Y^{3+}, Ba^{2+}, and Cu^{2+} complexes with polyacrylic [96] or polymethacrylic [97, 98] acid. More homogeneous distribution of metal ions in the polymer chain can be achieved by entering of them into a monomer molecule; such approach can be realized by means of a reaction of copolymerization of corresponding unsaturated metal carboxylates. For this purpose, for example, Y^{3+}, Ba^{2+}, and Cu^{2+} acrylates were mixed in 1:2:3 molar proportions in minimum quantity of methanol with its subsequent evaporation, and then solid-phase copolymerization was carried out [99, 100]. In another version of this method, single-phase stoichiometric films of HTSC ceramics are obtained by spray pyrolysis of solutions of the corresponding methacrylates (aerosol drops are applied onto a heated substrate; the precipitate is dried and twice annealed in an oxygen atmosphere, at 773 K and 1193 K) [101, 102]. This HTSC ceramic has a superconducting transition temperature of 87–92 K and a critical current density of up to 540 A cm^{-2}. Ferroelectric $PbTi_{0.6}Zr_{0.55}O_3$ films were obtained using aerosol deposition and pyrolysis of solutions of saturated metal carboxylates in methacrylic acid [103]. Transacylation (replacement of saturated carboxylate ligands in the metal coordination sphere by methacrylates) decreases the annealing temperature (to 923 K) and essentially improves the film surface quality: no pores or cracks are formed. Sol–gel synthesis also reduces the annealing temperature (to 823 K) in the preparation of ultradisperse layered $LiMO_2$ oxides (M=Co, Ni, V) (particle size 30–50 nm, specific surface area 2.3–17 m^2g^{-1}) in the presence of maleic [104] or polyacrylic [105] acids as chelating agents.

Perspective application field of the considered hybrid nanocomposites is alternative approaches for the immobilization of molecular magnets. It was shown that the magnetic $[Mn_{12}O_{12}(OOCCH=CH_2)_{16}$ cluster copolymerized with ethyl acrylate [106] or acrylic acid [107] in various molar ratios (25–200 mol.) does not undergo aggregation and is distributed homogeneously in a polymeric matrix. The obtained hybrid polymers showed superparamagnetic behavior at temperature higher than 8 K. It is notable that the obtained hybrid materials have properties of molecular magnets with characteristics close to the initial clusters despite the rather low content of cluster units (0.1–1 mol.%). So, field dependences of magnetization for methacrylate derivative of the earlier-stated cluster and its copolymer with MMA/cluster=200 composition have hysteresis loop with coercive force 0.8 and 0.1 T, accordingly [108]. Thus, polymerization of magnetic clusters in the presence of organic monomers allows to obtain magnetic materials that can be processed as typical polymers but that keep properties of molecular magnets.

4.5 CONCLUSION

Macromolecular metal carboxylates possess practically unlimited possibilities in the molecular and supramolecular structural organization. Many properties of metal-containing polymers are related to the formation of ion aggregates and multiplets in ionomers as well as three-dimensional network polymers. Of great interest are hybrid supramolecular structures on the base of metal dicarboxylates. It is reasonable to expect considerable advances in synthetic strategies of intercalation chemistry with respect to acrylate ions followed by polymerization in interlayered spaces of highly organized inorganic medium, liquid crystals, micelles, bilayer lipids, and so on. The various ways are found to transform isolated metal ions in polymer acids into metal nanoparticles. At the same time, a polycarboxylate matrix serves as stabilizing agent preventing aggregation of nanoparticles formed and ensures their homogeneous distribution in the composite. Metallopolymer nanocomposites have potential applications in fabricating devices with optical, electrical, and magnetic properties.

ACKNOWLEDGMENTS

This work is supported by the Russian Foundation for Basic Researches (project 13-03-92693) and by the program of fundamental researches of the Russian Academy Sciences Presidium number 24 "Basics of fundamental researches of nanotechnology and nanomaterials."

REFERENCES

[1] C. He, S.J. Lippard. *J. Am. Chem. Soc. 120*, 105 (**1998**).

[2] W. Ruttinger, G.C. Dismukes. *Chem. Rev. 97*, 1 (**1997**).

[3] A.D. Pomogailo, V.S. Savostyanov. *Synthesis and Polymerization of Metal-Containing Monomers* (CRC Press, Boca Raton, 1994).

[4] L. Rozes, N. Steunou, G. Fornasieri, C. Sanchez. *Monatsh. Chem. 137*, 501 (**2006**).

[5] D. Wöhrle, A.D.Pomogailo. *Metal Complexes and Metals in Macromolecules: Synthesis, Structures, and Properties.* (Wiley, Weinheim, 2003).

[6] S.M. Schlitter, H.P. Beck. *Chem. Ber. 129*, 1561 (**1996**).

[7] M.J. Vela, B.B. Snider, B.M. Foxman. *Chem. Mater. 10*, 3167 (**1998**).

[8] A. Galvan-Sanchez, F. Urena-Nunez, H. Flores-Llamas, R. Lorez-Castanares. *J. Appl. Polym. Sci. 74*, 995 (**1999**).

[9] J.H. O'Donnell, P.J. Pomery, R.D. Sothman. *Makromol. Chem. 181*, 409 (**1980**).

[10] M.R. Muidiniv, G.I. Dzhardimalieva, B.S. Selenova, A.D. Pomogailo. *Izv. Akad. Nauk. SSSR, Ser. Khim.* 2507 (**1988**).

[11] B.S. Selenova, G.I. Dzhardimalieva, M.V. Tsykalova, S.V. Kurmaz, V.P. Roshchupkin, I.Ya. Levitin, A.D. Pomogailo, M.E. Vol'pin. *Izv. Akad. Nauk. SSSR, Ser. Khim.* 500 (**1993**).

[12] A. Gbureck, U. Gbureck, W. Kiefer, U. Posset, R. Thull. *Appl. Spectrosc. 54*, 390 (**2000**).

[13] E.J. Ashford, V. Naldi, R. O'Dell, N.C. Billingham, S.P. Armes. *Chem. Commun.* 1285 (**1999**).

[14] V.L. Osborne, D.M. Jones, W.T.S. Huck. *Chem. Commun.* 1838 (**2002**).

[15] S. Tugulu, R. Barbey, M. Harms, M. Fricke, D. Volkmer, A. Rossi, H.-A. Klok. *Macromolecules 40*, 168 (**2007**).

[16] P. Molineux. *Water-Soluble Synthetic Polymers: Properties and Behavior* (CRC, Boca Raton, 1984).

[17] E.A. Bekturov, Z.B. Bakauova. *Synthetic Water-Soluble Polymers in Solution* (Huethig and Wepf, New York, 1986).

[18] A. Caetano, M.N. De Pinho, E.Drioli, H.Muntau (eds) *Membrane Technology: Application to Industrial Wastewater Treatment.* (Kluwer Academic, Dodrecht, 1995).

[19] I. Korus, M. Bodzek, K. Loska. *Sep. Sci. Technol. 17*, 111 (**1999**).

[20] K. Iwasa, I. Tasaki, R.C. Gibbons. *Science 210*, 338 (**1980**).

[21] I. Tasaki, P.M. Byrne. *Biopolymers 34*, 209 (**1994**).

[22] A. Eisenberg. *Macromolecules 3*, 147 (**1970**).

[23] A. Eisenberg, B. Hird, R. B. Moore. *Macromolecules 23*, 4098 (**1990**).

[24] D. Garcia, J.-S. Kim, A. Eisenberg. *J. Polym. Sci. B Polym. Phys. 36*, 2877 (**1998**).

[25] K. Suchocka-Galas. *Eur. Polym. J. 34* 127 (**1998**).

[26] S. Bagrodia, R. Pisipati, G.L. Wilkes, R.F. Storey, J.P. Kennedy. *J. Appl. Polym. Sci. 29*, 3065 (**1984**).

[27] C. Slusarczyk, A. Wlochowicz, A. Gronowski, Z. Wojtczak. *Polymer 29*, 1581 (**1988**).

[28] K. Suchocka-Galas, C. Slusarczyk, A. Wlochowicz. *Eur. Polym. J. 36*, 2175 (**2000**).

[29] K. Suchocka-Galas. *Eur. Polym. J. 25*, 1291 (**1989**).

[30] Y. Tsujita, M. Yasuda, M. Takei, T. Kinoshita, A. Takizawa, H. Yoshimizu. *Macromolecules 34*, 2220 (**2001**).

[31] S.-H. Kim, J.-S. Kim. *Macromolecules 36*, 1870 (**2003**).

[32] J.-S. Kim, Y.H. Nah, S.-S. Jarng. *Polymer 42*, 5567 (**2001**).

[33] S.-H. Kim, J.-S. Kim. *Macromolecules 36*, 2382 (**2003**).

[34] J.-S. Kim, G. Wu, A. Eisenberg. *Macromolecules 27*, 814 (**1994**).

[35] T. Kanamoto, I. Hatsua, S. Shirai, K. Tanaka. *Rep. Prog. Polym. Phys. Jpn. 16*, 245 (**1973**).

[36] K. Suchocka-Galas. *J. Appl. Polym. Sci. 96*, 268 (**2005**).

[37] H.-Q. Xie, W.-G. Yu, D. Xie, D.-T. Tian. *J. Macromol. Sci. A. Pure Appl. Chem. 44*, 849 (**2007**).

[38] R.T. Clay, R.E. Cohen. *Supramol. Sci. 2*, 183 (**1995**).

[39] G.I. Dzhardimalieva, V.A. Zhorin, I.N. Ivleva, A.D. Pomogailo, N.S. Enikolopyan. *Dokl. AN SSSR 287*, 654 (**1986**).

[40] G.I. Dzhardimalieva. *(Co)polymerization and thermal transformations as a way for synthesis of metallopolymers and nanocomposites. Doct. Sci. Chem. Thesis.* (ICPC RAS, Chernogolovka, 2009)

[41] S. Takamaizawa, M. Furihata, S. Takeda, K. Yamaguchi, W. Mori. *Macromolecules 33*, 6222 (**2000**).

[42] S. Takamaizawa, M. Furihata, S. Takeda, K. Yamaguchi, W. Mori. *Polym. Adv. Technol. 11*, 840 (**2000**).

[43] S. Takamaizawa, W. Mori, M. Furihata, S. Takeda, K. Yamaguchi. *Inorg. Chim. Acta. 283*, 268 (**1998**).

[44] F.A. Cotton, L.M. Daniels, C. Lin, C.A. Murillo, S-Y. Yu. *J. Chem. Soc. Dalton Trans. 502* (**2001**).

[45] S. Takamaizawa, T. Ohmura, K. Yamaguchi, W. Mori. *Molecular Cryst. Liquid Cryst. 342*, 199 (**2000**).

[46] M. Tanaka, I.Y. Park, K. Kuroda, C. Kato. *Bull. Chem. Soc. Jpn. 62*, 3442 (**1989**).

[47] S. Rey, J. Merida-Robles, K.S. Han, L. Guerlou-Demourgues, C. Delmas, E. Duguet. *Polym. Int. 48*, 277 (**1999**).

[48] C. Vaysse, L. Guerlou-Demourgues, E. Duguet, C. Delmas. *Inorg. Chem. 42*, 4559 (**2003**).

[49] C. Vaysse, L. Guerlou-Demourgues, C. Delmas, E. Duguet. *Macromolecules 37*, 45 (**2004**).

[50] K. Tajima, T. Aida. *Chem. Commun.* 2399 (**2000**).

[51] R.C. Smith, W.M. Fischer, D.L. Gin. *J. Am. Chem. Soc. 119*, 4092 (**1997**).

[52] H. Deng, D.L. Gin, R.C. Smith. *J. Am. Chem. Soc. 120*, 3522 (**1998**).

[53] D.H. Gray, D.L. Gin. *Chem. Mater. 10*, 1827 (**1998**).

[54] M.A. Reppy, D.H. Gray, B.A. Pindzola, J.L. Smithers, D.L. Gin. *J. Am. Chem. Soc. 123*, 363 (**2001**).

[55] S.R. Kline. *Langmuir 15*, 2726 (**1999**).

[56] A.D. Pomogailo, V.N. Kestelman. *Metallopolymer Nanocomposites* (Springer, Heidelberg, **2005**).

[57] M.Y. Gao, Y. Yang, B. Yang, F.L. Bian, J.C. Shen. *J. Chem. Soc. Chem. Commun.* 2779 (**1994**).

[58] Y. Wang, A. Suna, W. Mahler, R. Kasowski. *J. Chem. Phys. 87*, 7315 (**1987**).

[59] A.V. Volkov, M.A. Moskvina, I.V. Karachentsev, A.V. Rebrov, A.L. Volynskii, N.F. Bakeev. *Vysokomol. Soedin. A 40*, 45 (**1998**).

[60] A.S. Rozenberg, G.I. Dzhardimalieva, A.D. Pomogailo. *Polym. Adv. Technol. 9*, 527 (**1998**).

[61] A.D. Pomogailo, A.S. Rozenberg, G.I. Dzhardimalieva. *Controlled pyrolysis of metal-containing precursors as a way for syntheses of metallopolymer nanocomposites. "Metal-Polymer Nanocomposites"* ed. by. L. Nicolais, G. Carotenuto (Wiley, Hoboken, NJ, 2005, pp.75–122).

[62] A.S. Rozenberg, A.A. Rozenberg, G.I. Dzhardimalieva, A.D. Pomogailo. *Kolloid. Zh. 67*, 70 (**2005**).

[63] E. Sowka, M. Leonowicz, J. Kazmierczak, A. Slawska-Waniewska, A.D. Pomogailo, G.I. Dzhardimalieva. *Physica B Condens. Matter. 384*, 282 (**2006**).

[64] E.I. Aleksandrova, G.I. Dzhardimalieva, A.S. Rozenberg, A.D. Pomogailo. *Izv. Akad. Nauk. SSSR, Ser. Khim.* 308 (**1993**).

[65] M. Lawecka, A. Ślawska-Waniewska, K. Racka, M. Leonowicz, G.I. Dzhardimalieva, A.S. Rozenberg, A.D. Pomogailo. *J. Alloys Comp. 369*, 244 (**2004**).

[66] A.S. Rozenberg, E.I. Aleksandrova, G.I. Dzhardimalieva, A.N. Titkov, А.Н. Титков, A.D. Pomogailo. *Izv. Akad. Nauk. SSSR, Ser. Khim.* 1743 (**1993**).

[67] A.S. Rozenberg, G.I. Dzhardimalieva, N.V. Chukanov, A.D. Pomogailo. *Kolloid. Zh. 67*, 57 (**2005**).

[68] A.S. Rozenberg, E.I. Aleksandrova, G.I. Dzhardimalieva, N.V. Kyr'yakov, P.E. Chizhov, V.I. Petinov, A.D. Pomogailo. *Izv. Akad. Nauk. SSSR, Ser. Khim.* 885 (**1995**).

[69] A.S. Rozenberg, E.I. Aleksandrova, N.P. Ivleva, G.I. Dzhardimalieva, A.V. Raevskii, O.I. Kolesova, I.E. Uflyand, A.D. Pomogailo. *Izv. Akad. Nauk. SSSR, Ser. Khim.* 265 (**1998**).

[70] A.T. Shuvaev, A.S. Rozenberg, G.I. Dzhardimalieva, N.P. Ivleva, V.G. Vlasenko, T.I. Nedoseikina, T.A. Lubeznova, I.E. Uflyand, A.D. Pomogailo. *Izv. Akad. Nauk. SSSR, Ser. Khim.* 1505 (**1998**).

[71] M. Lawecka, M. Kopcewicz, A. Slawska-Waniewska, M. Leonowicz, J. Kozubowski, G.I. Dzhardimalieva, A.S. Rozenberg, A.D. Pomogailo. *J. Nanopart. Res. 5*, 373 (**2003**).

[72] A.S. Rozenberg, G.I. Dzhardimalieva, A.D. Pomogailo. *Dokl. Akad. Nauk 356*, 66 (**1997**).

[73] A.D. Pomogailo, A.S. Rozenberg, G.I. Dzhardimalieva. *Russ. Chem. Rev. 80*, 257 (**2011**).

[74] S. Sun, C.B. Murray, D. Weller, L. Folks, A. Moser. *Science 287*, 1989 (**2000**).

[75] S.-H. Choi, E.-G. Kim, T. Hyeon. *J. Am. Chem. Soc. 128*, 2520 (**2006**).

[76] K. An, N. Lee, J. Park, S.C. Kim, Y. Hwang, J.G. Park, J.Y. Kim, J.H. Park, M.J. Han, J. Yu, T. Hyeon. *J. Am. Chem. Soc. 128*, 9753 (**2006**).

[77] T. Hyeon. *Chem. Commun.* 927 (**2003**).

[78] Li ChenSha, Li YuNing, Wu Yiliang, Beng S. Ong, Rafik O. Loutfy. *Sci. China E. Techn. Sci. 51*, 2075 (**2008**).

[79] S.-H. Choi, K. An, E.-G. Kim, J.H. Yu, J.H. Kim, T. Hyeon. *Adv. Funct. Mater. 19*, 1645 (**2009**).

[80] S.-H. Choi, E.-G. Kim, J. Park, K. An, N. Lee, C. Kim, T. Hyeon. *J. Phys. Chem. B 109*, 14792 (**2005**).

[81] M. Moffit, H. Vali, A. Eisenberg. *Chem. Mater. 10*, 1021 (**1998**).

[82] M. Moffit, L. McMahon, V. Pessel, A. Eisenberg. *Chem. Mater. 7*, 1185 (**1995**).

[83] N. Duxin, F.Liu, H. Vali, A. Eisenberg. *J. Am. Chem. Soc. 127*, 10063 (**2005**).

[84] S.A. Meenach, K.W. Anderson, J.Z. Hilt. *Hydrogel nanocomposites: biomedical applications, biocompatibility, and toxicity analysis. Chapter in Book "Safety of Nanoparticles, Nanostructure Science and Technology"* ed. by T.J. Webster (Springer, Heidelberg, 2009).

[85] S. Laurent, D. Forge, M. Port, A. Roch, C. Robic, L.V. Elst, R.N. Muller. *Chem. Rev. 108*, 2064 (**2008**).

[86] J. Zhang, S. Xu, E. Kumacheva. *J. Am. Chem. Soc. 126*, 7908 (**2004**).

[87] C. Bai, Y. Fang, Y. Zhang, B. Chen. *Langmuir 20*, 263 (**2004**).

[88] G. Kickelbick. *Prog. Polym. Sci. 28*, 83 (**2003**).

[89] U. Schubert. *J. Sol–gel Sci. Technol. 26*, 47 (**2003**).

[90] F. Ribot, F. Banse, C. Sanchez, M. Lahcini, B. Jousseaume. *J. Sol–Gel Sci. Technol. 8*, 529 (**1997**).

[91] S. Gross, A. Zattin, V.D. Noto, S. Lavina. *Monatsh. Chem. 137*, 583 (**2006**).

[92] L. Armelao, C. Eisenmenger-Sittner, M. Groenewolt, S. Gross, C. Sada, U. Schubert, E. Tondello, A. Zattin. *J. Mater. Chem. 15*, 1838 (**2005**).

[93] X.M. Liu, S.-Y. Fu. *J. Magn. Magn. Mater. 308*, 61 (**2007**).

[94] X.M. Liu, S.Y. Fu, H.M. Xiao, C.J. Huang. *Physica B 370*, 14 (**2005**).

[95] H.-M. Xiao, X.-M. Liu, S.-Y. Fu. *Compos. Sci. Technol. 66*, 2003 (**2006**).

[96] S. Dubinsky, Y. Lumelsky, G.S. Grader, G.E. Shter. *J. Polym. Sci. B. Polym. Phys. 43*, 1168 (**2005**).

[97] J.C.W. Chien, B.M. Gong, J.M. Madsen, R.B. Hallock. *Phys. Rev. B 38*, 11853 (**1988**).

[98] J.C.W. Chien, B.M. Gong, X. Mu, Y. Yang. *J. Polym. Sci. Polym. Chem. Ed. 28*, 1999 (**1990**).

[99] A.D. Pomogailo, V.S. Savostyanov, G.I. Dzhardimalieva, A.V. Dubovitskii, A.N. Ponomarev. *Izv. Akad. Nauk. SSSR, Ser. Khim.* 1096 (**1995**).

[100] V.S. Savostyanov, V.A. Zhorin, G.I. Dzhardimalieva, A.D. Pomogailo, A.V. Dubovitskii, V.N. Topnikov, M.K. Makova, A.N. Ponomarev. *Dokl. Akad. Nauk 318*, 378 (**1991**).

[101] Yu A. Tomashpolskii, L.F. Rybakova, O.F. Fedoseeva, I.A. Noskova, S.A. Menshykh. *Neorg. Mater. 37*, 75 (**2001**).

[102] T.A. Starostina, O.P. Syutkina, L.F. Rybakova, V.V. Bogatko, R.R. Shifrina, Yu. N. Venevtsev. *Zh. Neorg, Khim. 37*, 2402 (**1992**).

[103] Yu A. Tomashpolskii, L.F. Rybakova, T.V. Lunina, O.F. Fedoseeva, S.G. Prutchenko, S.A. Menshykh. *Neorg. Mater. 37*, 596 (**2001**).

[104] I.-H. Oh, S.-A. Hong, Y.-K. Sun. *J. Mater. Sci. 32*, 3177 (**1997**)

[105] Y.-K. Sun, I.-H. Oh, S.-A. Hong. *J. Mater. Sci. 31*, 3617 (**1996**).

[106] F. Palacio, P. Oliete, U. Schubert, I. Mijatovic, N. Husing, H. Peterlik. *J. Mater. Chem. 14*, 1873 (**2004**).

[107] R.Cusnir, G. Dzhardimalieva, S. Shova, D.Prodius, N. Golubeva, A. Pomogailo, C. Turta. Proceed. Int. Conference on Coordination Chemistry-ICCC38, Jerusalem, Israel, 20–25 July, **2008**, p. 475.

[108] S. Willemin, B. Donnadien, L. Lecren, B. Henner, R. Clerac, C. Guerin, A.V. Pokrovskii, J. Larionova. *New J. Chem. 28*, 919 (**2004**).

5

IN-SITU MICROWAVE-ASSISTED FABRICATION OF POLYMERIC NANOCOMPOSITES

H. SadAbadi, S. Badilescu,
M. Packirisamy, and R. Wüthrich

*Optical Bio-Micro Systems Laboratory, Department of Mechanical Engineering,
Concordia University, Montreal, QC, Canada*

5.1 INTRODUCTION

Metal–polymer nanocomposite or nanometal–polymer composite films are hybrid materials with nanometer-sized inorganic nanoparticles immobilized, uniformly dispersed, and integrated into a polymer matrix. The scientific and technological interest in these hybrid materials, in particular, those based on Au and Ag but also on semiconductors such as TiO_2 or ZnO, stems mostly from their optical properties. The inorganic nanoparticles used to enhance the performance of the polymer, especially their optical and mechanical properties, are called "optically effective additives," and they lead to new functionalities of the polymer-based materials, without the loss of their transparency [1, 2]. Nanocomposites may show optical nonlinearities and/or ultralow or ultrahigh refractive indices and are suitable for applications such as color filters, optical sensors, data storages, waveguides, optical strain detectors, and thermochromic materials. Other applications of gold and silver polymer and copolymer composites, for example, for water purification, targeted drug release, and antimicrobial coatings, have been reported as well [3].

Nanocomposites: In Situ *Synthesis of Polymer-Embedded Nanostructures*, First Edition.
Edited by Luigi Nicolais and Gianfranco Carotenuto.
© 2014 John Wiley & Sons, Inc. Published 2014 by John Wiley & Sons, Inc.

Nanocomposites can be synthesized through different approaches, basically, either by *in situ* methods or by incorporating premade nanoparticles into a polymer matrix by using a common solvent. *In situ* methods are beneficial to avoid the aggregation of inorganic particles inside the polymer matrix, leading to isolated primary particles. Physical methods such as chemical vapor deposition, ion implantation, and thermolysis have also been used successfully [4, 5]. For a description of recent developments in the synthesis and applications of nanocomposite materials, see some excellent reviews and the references herein [6–11].

Gold – poly(dimethylsiloxane) (Au–PDMS) and silver–poly(dimethylsiloxane) (Ag–PDMS) nanocomposites are of considerable interest due to the simplicity of their preparation, low cost, good transparency, oxidative stability, and nontoxicity to cells. In addition, PDMS, the host polymer, has a low glass transition temperature (T_g), excellent flexibility, high thermal and oxidative stability, and good hemo- and biocompatibility. A major drawback is the strong hydrophobicity and, therefore, the inertness to biological molecules. To render the PDMS surface hydrophilic, functional groups have been introduced by low-pressure plasma treatments, corona discharge treatments, etc., to promote the filling of the microchannels with aqueous solutions and to facilitate PDMS microchip bonding as well [12–14]. However, while freshly modified surfaces show good or moderate wettability, this effect is not stable and the hydrophobicity is regained over the time. The hydrophobic recovery is usually accounted for by the migration of the low-molecular-weight constituents from the bulk to the polymer surface. Gold – poly(dimethylsiloxane) nanocomposites have been synthesized in the form of gels, foams, and films for applications including water purification and drug delivery [15]. Because of their strong plasmon band in the visible (Vis) spectrum, arising from the excitation of the plasmon by the incident light, gold nanoparticles are particularly suited for sensing applications. Due to the wide utilization of gold nanoparticles with various shapes for biosensing, biological labeling, etc., their association with PDMS, in the form of nanocomposite materials, opens new possibilities in microfluidic biosensing. We, and other groups, have prepared Au–PDMS nanocomposites by *in situ* nanoparticle formation methods and obtained well-dispersed nanoparticles with a narrow size distribution [16–18]. However, the biosensing properties of the as-prepared nanocomposite films have been found to be poor. We have shown in a previous paper that the sensitivity of the Ag–PDMS platforms can be increased by functionalization, followed by a post-synthesis thermal treatment at moderate temperatures, treatment inducing an important morphological change [19].

Generally, the sensing properties of nanocomposites depend on quite a few parameters related to the conditions of their preparation, that is, the nature of the precursor molecule, the concentration of the solutions, the temperature, and the thickness of the film. They are determinant for the distribution of the metal particles in the polymer matrix. On the other hand, and this is the most important issue, the polymer has to provide an appropriate environment for the metal nanoparticles in order to be able to interact with the biomolecules [20]. Because of its extreme hydrophobicity, PDMS does not promote the hydrophilic biomolecules to enter the inner domains of the polymer. However, it has been reported recently that by varying the concentration of the cross-linker, the free volume of the film can be changed and the enclosure of various molecular entities can be facilitated [21].

It is well established that the decay length of the electromagnetic field for LSPR sensors depends on the size, shape, and composition of nanoparticles and that it is not more than 5–15 nm, permitting the detection of only very thin layers of adsorbate molecules [22–25]. However, if the gold nanoparticles are embedded deep into the polymer, the biomolecules may be too far from the sensing volume of the particle. In addition, if the interaction between the particles and the polymer chains is too strong and the nanoparticles are entangled in the polymer network, their mobility may be severely restricted. Since the biosensing properties are determined by the spatial distribution of Au nanoparticles, it is important to be able to control this distribution. We have explored some promising ways to increase the mobility of Au nanoparticles in order to concentrate most of them on the surface of the film and, thus, make them accessible to the surrounding environment containing the biomolecules of interest. Among these methods is the thermal post-synthesis treatment of Au–PDMS nanocomposite that has been proved to be an adequate method to improve the distribution of Au nanoparticles in polymers. In our work, Au–PDMS nanocomposite materials are intended to be further used for microcantilever sensing, and therefore, a high degree of elasticity is required. For this reason, a lower concentration of cross-linking agent has to be used, and the resulting low free volume of the polymer does not favor the introduction of a high amount of gold nanoparticles. Swelling experiments have been thought to be useful to expand the polymer network and allow aqueous solvents to infiltrate the material [26].

In this work, we are reporting our results on a novel *in situ* synthesis of Au–PDMS by using a microwave-induced reduction of gold ions. The microwave heating is fundamentally different from the conventional one. In the microwave heating, molecular dipoles are induced to oscillate, and this oscillation causes a higher rate of molecular collisions, which, in turn, generates heat. Since microwave irradiation is a volumetric phenomenon, heat is generated in the irradiated volume with a homogenous heat distribution. Microwave heating has been studied as a promising technique for nanoparticle synthesis because it generates immediately nucleation sites in the solution, which considerably increases the rate of reactions. Due to the rapid internal heating, the kinetics of the reaction is much improved. Compared with the conventional methods, microwave synthesis of nanoparticles has the advantages of short reaction time, small particle size, narrow particle size distribution, and high purity. Microwave irradiation has been used for the preparation of gold and silver nanoparticles under various conditions [27, 28], but the *in situ* synthesis in a polymer matrix has not been yet explored. In the present work, we have investigated the microwave-induced reduction of gold ions onto a PDMS matrix and the usefulness of the nanocomposite for sensing application.

5.2 EXPERIMENTAL

5.2.1 Materials

The Sylgard® 184 elastomer kit for the PDMS fabrication is purchased from Dow Corning Corporation. Gold chloride trihydrate ($HAuCl_4 \cdot 3H_2O$) from Alfa Aesar and 1-pentanol, toluene, N, N-dimethylformamide, 2-propanol, and ethyl alcohol are purchased from Sigma-Aldrich.

5.2.2 Fabrication of PDMS

For the fabrication of PDMS, the base polymer and the curing agent were mixed by using a 10:1 ratio (by weight). The fabrication is followed by the degasification of the mixture in a vacuum desiccator for 10 min. In order to obtain a very smooth surface, a silanized silicon wafer is used as a mold. The mixture was then poured in the flat mold, and after a second degasifying process (10 min), the mold is kept in oven at 70 °C overnight for curing. Afterward, the cured PDMS was peeled off from the mold and cut in 1 cm × 4 cm samples. The thickness of the PDMS sample was around 1 mm.

5.2.3 Synthesis of Au–PDMS Nanocomposite

Solutions of chloroauric acid in ethanol are prepared with concentrations ranging from 0.1% to 5%. The fabrication of the nanocomposite starts with dropping 60 μL of the ethanol solution of the chloroauric acid on a PDMS sample. Then the solution is spread over the whole area after keeping it for at least 10 min in contact with the PDMS surface. The samples are then introduced into a microwave oven and irradiated for different times. Figure 5.1 shows the procedure and some of the samples prepared by using this procedure. To study the characteristics of the gold film formed on the PDMS surface, UV/Vis spectroscopy and scanning electron microscopy (SEM) imaging is employed.

5.2.4 Annealing Process

To improve the properties of the Au–PDMS network for biosensing applications, the samples were annealed. Heat treatment has been applied to increase the concentration of Au nanoparticles on the surface of the sample. To do this, the samples prepared using the *in situ* method described earlier have been heated gradually, from room temperature to 300°–350 °C, and then kept at the maximum temperature for 30 min.

5.2.5 Tension Test

The tension test has been performed by using BOSE ElectroForce® 3200 test instruments. The testing was performed on both the as-prepared and the annealed samples, respectively. The stress–strain graph for each sample is obtained, and the slope of the graph corresponding to the module of elasticity is calculated.

5.2.6 Sensitivity Test

The sensitivity of the nanocomposite to the dielectric media with different refractive indices has been performed. For this test, six solvents have been selected and their refractive indices are given in Table 5.1. The sensitivity has been calculated from the

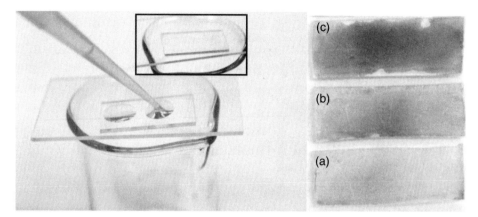

Figure 5.1. Procedure for the preparation. The ethanol solution of the gold salt (1%) is dropped on the PDMS surface. The inset shows the solution expanded on the sample's surface before irradiation. The samples irradiated for different times varying from (**a**) 30 s to (**b**) 60 s and (**c**) 90 s. (*See insert for color representation of the figure.*)

TABLE 5.1. Refractive Index of Solvents Used for Sensitivity Test

Environment	Refractive index (n)
Deionized water	1.33
Ethyl alcohol	1.36
2-propanol	1.38
1-pentanol	1.41
N, N-dimethylformamide	1.43
Toluene	1.50

plot $\Delta\lambda_{max} - \Delta n$ where $\Delta\lambda_{max}$ is the Au LSPR band shift (in nm) and Δn is the refractive index difference between the solvent and DI water.

5.3 RESULTS AND DISCUSSION

5.3.1 Microwave-Induced Reduction of Gold Ions

The first step in the fabrication of the nanocomposite is spreading of the gold solution on the surface before irradiation. However, because of the strong hydrophobicity of the PDMS surface, the solution can be spread only after minimum 10 min of contact between the solution and the PDMS surface.

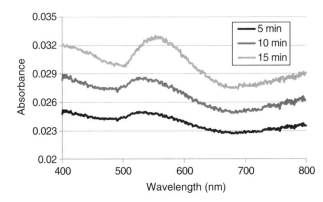

Figure 5.2. Seed formation on the surface of PDMS. (*See insert for color representation of the figure.*)

To investigate the interaction between the solution and the polymer surface, samples were prepared by keeping the solution on the PDMS surface for varying times (5–15 min). Afterward, the samples were washed and their spectra were recorded (Figure 5.2). The spectra show the presence of Au band as an evidence of gold seeds on the surface. However, the samples are not red, and the SEM images of the sample do not show the presence of particles, probably because the particles are in the range of sub-nanometer, which is smaller than the resolution of the SEM instrument.

From these results, it can be inferred that the formation of the gold seeds on the PDMS surface alters the surface properties by reducing the hydrophobicity and allows the spreading of the solution. In addition, the seeds, most probably, initiate the formation of the gold nanoparticles on the surface during the irradiation.

The experiments indicated that by keeping the irradiated samples in the atmosphere, the color of the samples is changing with time. Indeed, immediately after irradiation, the samples appear gray yellowish, but after keeping them in the room temperature for an hour, the color turns to red. Figure 5.3 illustrates the change of color for a sample irradiated for 90 s, and the spectra (Figure 5.4) show that the position of the Au LSPR band shifts to shorter wavelengths during the process. Leaving the sample for 75 min at room temperature, the Au band shifts from 552 nm to 531 nm.

To accelerate the formation of gold nanoparticles onto the surface of PDMS, instead of keeping the samples for 75 min in the atmosphere, they can be immersed in ethanol immediately after MW irradiation and kept for at least 1 min. The spectra show that in both cases the resulting nanocomposites are identical.

A possible explanation of the aforementioned results can be the following: the microwave irradiation may have two effects—the changes of the surface properties by inducing the formation of polar groups on the surface. At the same time, it may change the distribution of the curing agent in the polymer, bringing a large amount on the surface. This results in the formation of a high density of gold aggregates on the surface (gray-yellowish color, Figure 5.3b). Upon immersing the sample in ethanol, a part of the aggregates are carried by ethanol into the polymer network, and consequently, the color

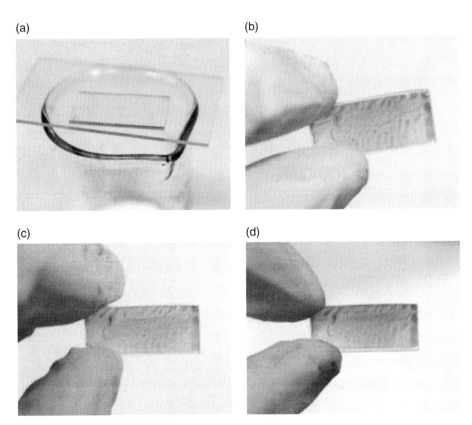

Figure 5.3. (a) Nonirradiated sample; the samples corresponding to different times of irradiation: (b) immediately after MW irradiation (90 s), (c) 30 min after MW irradiation, (d) 75 min after MW irradiation. The color is stabilized afterward. (*See insert for color representation of the figure.*)

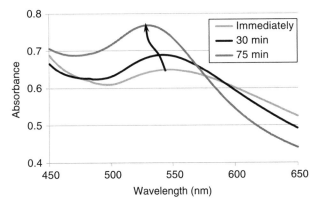

Figure 5.4. Spectra of the samples kept in the atmosphere for different times after irradiation. (*See insert for color representation of the figure.*)

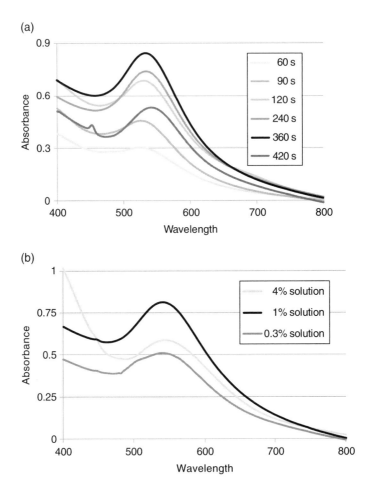

<u>Figure 5.5.</u> (a) Effect of the irradiation time and (b) of the concentration of gold salt in solution on the spectra of the samples. (*See insert for color representation of the figure.*)

turns to red because more individual gold nanoparticles will remain on the surface (Figure 5.3d). If the gray-yellowish sample is kept in the atmosphere, the same process will happen but at a much slower rate (75 min).

The morphology of the samples prepared through the reduction of gold ions on the surface of PDMS depends on both the irradiation time and the concentration of the precursor solution (Figure 5.5). From the spectra, it can be inferred that a 1% solution of gold salt, irradiated for 360 s, results in a sample with a narrow Au LSPR band that corresponds to a narrow size distribution.

The SEM results (not given here) show that for a low irradiation time (less than 15 s), there are no nanometer-sized particles on the surface. By increasing the irradiation time, for example, for 50 s, more nanoparticles begin to appear on the surface and their sizes are in the range of 5 ± 1 nm. For 90 s irradiation, the particle size reaches 25 ± 5 nm. Upon further increasing the time, linear aggregates can be seen on the surface of the sample.

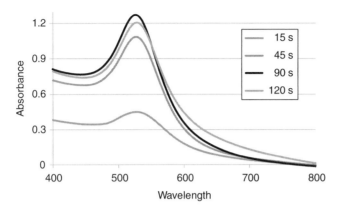

Figure 5.6. Spectra of the annealed samples corresponding to different irradiation times. (*See insert for color representation of the figure.*)

The formation of gold nanoparticles on the surface is accompanied by the penetration of the solution into the polymer network and reduction of the gold salt by the crossing agent present in PDMS. The spectra shown in Figure 5.5 display a very broad band due to the wide size distribution. To improve the distribution of the gold nanoparticles embedded in the surface of PDMS, a heat treatment has been employed.

Samples prepared with different times of irradiation by using the 1% solution of gold salt have been annealed, and the corresponding Au LSPR bands are shown in Figure 5.6. The figure indicates that in order to prepare a nanocomposite with a narrow size distribution of gold nanoparticles, a 1% solution of gold precursor has to be spread and irradiated 90 s on the surface of PDMS. The reaction between the gold ions and curing agent is the following [17]:

$$3\,\overset{|}{\underset{|}{Si}} - H + 3/2 H_2O + 2 AuCl_4^- \rightarrow 3/2 - \overset{|}{Si} - O - \overset{|}{Si} - 2Au + 8Cl^- + 6H^+$$

It should be noted that in all cases, the samples are kept at least 75 min at room temperature before the annealing or immersed in ethanol, to stabilize the color of the samples, as mentioned earlier in this section.

Scanning electron microscopy images have been used to show the gold nanoparticles embedded onto the PDMS surface. The images before and after annealing, respectively, are given in Figure 5.7. The figure shows that after annealing at high temperatures, the aggregated particles are transformed into individual particles uniformly distributed on the surface. However, gold nanoparticles are distributed inside the polymer matrix as well.

The spectra corresponding to the synthesized and annealed, respectively, are given in Figure 5.8 and show that, generally after annealing, the Au plasmon band shifts to shorter wavelengths (blue shift). The blue shift after annealing reflects the improved morphology, in terms of uniformity of the nanoparticles distribution as shown in the SEM images in Figure 5.7b.

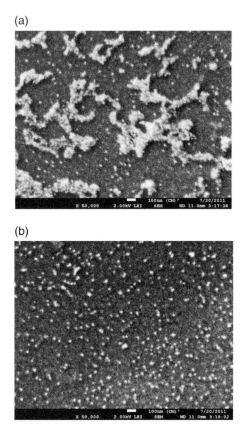

Figure 5.7. Scanning electron microscopy images of Au–PDMS nanocomposite film, (a) as-prepared sample (b) annealed sample at 300–350 °C.

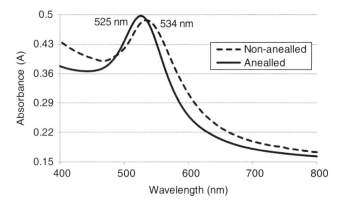

Figure 5.8. Effect of the annealing on the position of the Au LSPR band.

TABLE 5.2. Module of Elasticity for PDMS with Two Different Ratios (MPa)

Sample type	PDMS (10:1)	PDMS (4:1)
As-prepared	1.27 MPa	0.74 MPa
Post-processed (A)*	1.31 MPa	0.87 MPa

*A, annealed

TABLE 5.3. Effect of Heat Treatment of the Module of Elasticity

Sample type	PDMS	Au–PDMS nanocomposite
As-prepared	1.27 ± 0.03 MPa	1.47 ± 0.1 MPa
Post-processed (A)*	1.31 ± 0.05 MPa	1.52 ± 0.1 MPa

*A, annealed

5.3.2 Mechanical Properties

To investigate the effect of the amount of the curing agent on the elasticity of PDMS, tension tests have been performed on samples prepared with two different ratios. The ratios of 10:1 and 4:1 are selected and the results are shown in Table 5.2. The results show that by using a higher ratio, samples with a higher elasticity are obtained. For each case, four similar samples were tested and the average values are shown in the table.

The tension test is also used to show how the presence of gold nanoparticles affects the mechanical behavior of the samples. Samples (fabricated as described in section 5.2.2) were irradiated 90 s by using the 1% solution of the gold salt. Four identical samples were tested for both as-prepared and post-processed nanocomposites, and the results are given in Table 5.3. The results show that the presence of gold nanoparticles increases the module of elasticity up to 20%. In addition, the high elasticity of the annealed sample indicates the existence of gold nanoparticles inside of the polymer network, even after the annealing process.

5.3.3 Effect of Annealing on the Sensitivity

By increasing the refractive index of the surrounding medium, the Au band is shifted to longer wavelengths. The sensitivity graph for the as-prepared and annealed samples, respectively, is depicted in Figure 5.9.

The results show that the annealed sample has a sensitivity as high as 77 nm RIU^{-1}, which makes it useful for label-free biosensing of proteins. In addition, the regression coefficient (R) is given for each graph to show the discrepancy of the data points. The R value shows that in the case of the annealed sample, the data follow better the trend than in the case of as-prepared platform.

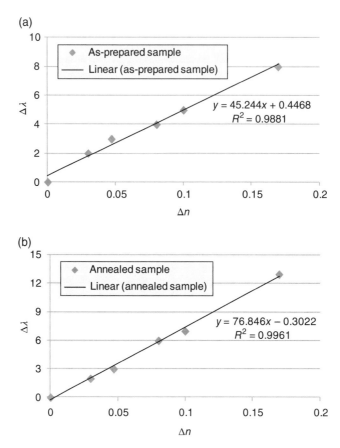

<u>Figure 5.9.</u> Sensitivity of nanocomposite sensing platforms (the nanocomposite is prepared by using a 1% gold precursor solution with an irradiated time of 90 s) **(a)** as-prepared **(b)** annealed.

5.4 CONCLUSION

Gold – poly(dimethylsiloxane) nanocomposite films were prepared through a novel microwave-induced reaction. In this method, the gold precursor (chloroauric acid) is dissolved in ethanol, and due to high permeability of the carrier fluid in the PDMS network, the rate of the reduction of gold is increased.

It is found that during the interaction between PDMS and the solution, sub-nanometer-sized gold seeds are formed on the surface, which possibly initiate the formation of gold nanoparticles during the irradiation. The gold nanoparticles prepared by microwave irradiation are distributed in the polymer matrix, and to increase the density of nanoparticles on the surface, an annealing process is employed.

The mechanical tension test shows that the presence of nanoparticles increases the module of elasticity of PDMS up to 20%. In addition, a sensitivity of 77 nm RIU^{-1} is obtained in the case of the annealed sample. The proposed platform can be used effectively for label-free biosensing experiments. The main advantages of the method are the simplicity as well as the increased rate of the reaction. The narrow size distribution of the synthesized particles obtained here is relevant for microfluidic sensing.

REFERENCES

[1] W. Caseri, "Inorganic nanoparticles as optically effective additives for polymers", *Chem. Eng. Commun.* **196**, 549–572 (2009).

[2] W. Caseri, "Nanocomposites of polymers and metals or semiconductors: historical background and optical properties", *Macromol. Rapid Commun.* **21**, 705–722 (2000).

[3] F.S. Aggor, E.M. Ahmed, A.T. El-Aref, M.A. Asem, "Synthesis and characterization of poly(acrylamide-co-acrylic acid) hydrogel containing silver nanoparticles for antimicrobial applications", *J. Am. Sci.* **6**, 648–656 (2010).

[4] A.L. Stepanov, R.I. Khaibullin, "Optics of metal nanoparticles fabricated in organic matrix by ion implantation", *Rev. Adv. Mater. Sci.* **7**, 108–125 (2004).

[5] G. Carotenuto, G. La Peruta, L. Nicolais, "Thermo-chromic materials based on polymer-embedded silver clusters", *Sens. Actuators. B Chem.* **114**, 1092–1095 (2006).

[6] M. Vidotti, R.F. Carvalhal, R.K. Mendes, D.C.M. Ferreira, L.T. Kubota, "Biosensors based on gold nanostructures", *J. Braz. Chem. Soc.* **22**, 3–20 (2011).

[7] P.H. Cury Camargo, K.G. Satyanarayana, F. Wypych, "Nanocomposites: Synthesis, structure, properties and new application opportunities", *Mater. Res.* **12**, 1–39 (2009).

[8] D. Li, C. Li, A. Wang, Q. He, J. Li, "Hierarchical gold/copolymer nanostructures as hydrophobic nanotank for drug encapsulation", *J. Mater. Chem.* **20**, 7782–7787 (2010).

[9] L.L. Beecroft, C.K. Ober, "Nanocomposite materials for optical applications", *Chem. Mater.* **9**, 1302–1317 (1997).

[10] H. Althues, J. Henle, S. Kaskel, "Functional inorganic nanofillers for transparent polymers", *Chem. Soc. Rev.* **36**, 1454–1465 (2007).

[11] S. Li, M. Meng Lin, M.S. Toprak, D.K. Kim, M. Muhammed, "Nanocomposites of polymer and inorganic nanoparticles for optical and magnetic applications", *Nano Rev.* **1**, 1–19 (2010).

[12] S.K. Sia, G.M. Whitesides, "Microfluidic devices fabricated in poly(dimethylsiloxane) for biological studies", *Electrophoresis* **24**, 3563–3576 (2003).

[13] I. Wong, C.-M. Ho, "Surface molecular property modifications for poly(dimethylsiloxane) (PDMS) based microfluidic devices," *Microfluid. Nanofluidics* **7**, 291–306 (2009).

[14] J.A. Vickers, M.M. Caulum, C.S. Henry, "Generation of Hydrophilic poly(dimethylsiloxane) for high-performance microchip electrophoresis", *Anal. Chem.* **78**, 7446–7452 (2006).

[15] A. Scott, R. Gupta, G.U. Kulkarni, "A simple water-based synthesis of Au nano-particle/ PDMS composites for water purification and targeted drug release", *Macromol. Chem. Phys.* **211**, 1640–1647 (2010).

[16] P. Devi, A.Y. Mahmoud, S. Badilescu, M. Packirisamy, P. Jeevanandam, V.-V. Truong, "Synthesis and Surface Modification of Poly (dimethylsiloxane) – Gold Nanocomposite Films for Biosensing Applications", *In Proceeding of Biotechno Conference, Cancun, Mexico*, 1–5 (2010).

[17] Q. Zhang, J.-J. Xu, Y. Liu, H.-Y. Chen., "In-situ synthesis of poly(dimethylsiloxane)-gold nanoparticle composite films and its application in microfluidic systems", *Lab Chip* **8**, 352–357 (2008).

[18] A. Goyal, A. Kumar, P.K. Patra, S. Mahendra, S. Tabatabaei, P.J.J. Alvarez, G. John, P. M. Ajayan, "In situ synthesis of metal nanoparticle embedded free standing multifunctional PDMS films", *Macromol. Rapid Commun.* **30**, 1116–1122 (2009).

[19] J. Ozhikandathil, S. Badilescu, M. Packirisamy, "Synthesis and characterization of silver-PDMS nanocomposite for biosensing applications", *In Proceeding of Photonic North Conference, Ottawa, Canada*, (2011).

[20] R. Gradess, R. Abarguez, A. Habbou, J. Canet-Ferrer, E. Pedrueza, A. Russell, J.L. Valdés, J.P. Msrtinez-Pastor, "Localized surface plasmon resonance sensor based on Ag-PVA nanocomposite films", *J. Mater. Chem.* **19**, 9233–9240 (2009).

[21] M. Tagaya, M. Nakagawa, "Incorporation of decanethiol-passivated gold nanoparticles into cross-linked poly (dimethyl siloxane) films", *Smart Mater. Res.*, 7p (2011).

[22] G. Barbillon, J.-L. Bijeon, J. Plain, M. Lamy de la Chapelle, P.-M. Adam, P. Royer, "Biological and chemical nanosensors based on localized surface plasmon resonance", *Gold Bull.* **40**, 240–244 (2007).

[23] D.A. Stuart, A.J. Haes, C.R. Yonzon, E.M. Hicks, R.P. Van Duyne, "Biological applications of localized surface plasmonic phenomenae", *IEEE Proc.-Nanobiotechnol.* **152**, 13–32 (2005).

[24] A.J. Haes, S. Zou, G.C. Schatz, R.P. Van Duyne, "Nanoscale optical biosensor: short range distance dependence of the localized surface Plasmon resonance of noble metal nanoparticles", *J. Phys. Chem. B* **108**, 6961–6968 (2004).

[25] A.J. Haes, S. Zou, G.C. Schatz, R.P. Van Duyne, "A nanoscale optical biosensor: the long range distance dependence of the localized surface plasmon resonance of noble metal nanoparticles", *J. Phys. Chem. B* **108**, 109–116 (2004).

[26] H. SadAbadi, S. Badilescu, M. Packirisamy, R. Wuthrich, "PDMS-gold nanocomposite platforms with enhanced sensing properties", *J. Biomed. Nanotechnol.* **8**, 539–549 (2012).

[27] Ming Shen, Yukou Du, Nanping Hua, Ping Yang, "Microwave irradiation synthesis and self-assembly of alkylamine-stabilized gold nanoparticles", *Powder Technol.* **162**, 64–72 (2006).

[28] B. Aswathy, G.S. Avadhani, I.S. Sumithra, S. Suji, G. Sony, "Microwave assisted synthesis and UV–Vis spectroscopic studies of silver nanoparticles synthesized using vanillin as a reducing agent", *J. Mol. Liq.* **159**, 165–169 (2011).

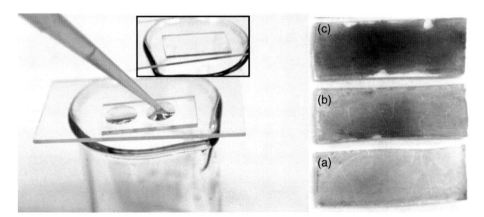

Figure 5.1. Procedure for the preparation. The ethanol solution of the gold salt (1%) is dropped on the PDMS surface. The inset shows the solution expanded on the sample's surface before irradiation. The samples irradiated for different times varying from (a) 30 s to (b) 60 s and (c) 90 s.

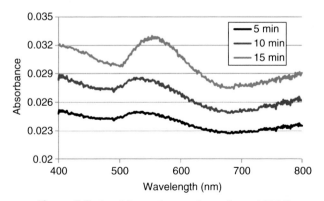

Figure 5.2. Seed formation on the surface of PDMS.

Nanocomposites: In Situ *Synthesis of Polymer-Embedded Nanostructures*, First Edition.
Edited by Luigi Nicolais and Gianfranco Carotenuto.
© 2014 John Wiley & Sons, Inc. Published 2014 by John Wiley & Sons, Inc.

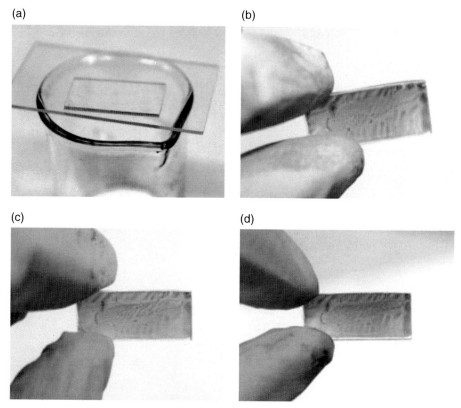

Figure 5.3. (a) Nonirradiated sample; the samples corresponding to different times of irradiation: (b) immediately after MW irradiation (90 s), (c) 30 min after MW irradiation, (d) 75 min after MW irradiation. The color is stabilized afterward.

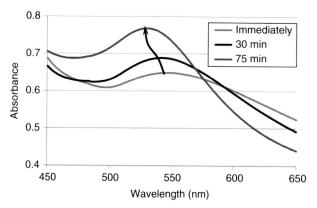

Figure 5.4. Spectra of the samples kept in the atmosphere for different times after irradiation.

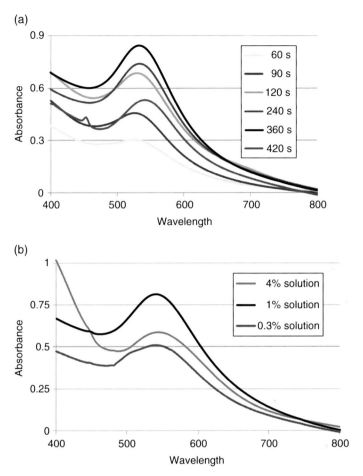

Figure 5.5. (a) Effect of the irradiation time and (b) of the concentration of gold salt in solution on the spectra of the samples.

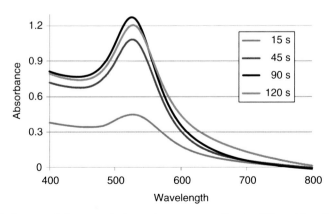

Figure 5.6. Spectra of the annealed samples corresponding to different irradiation times.

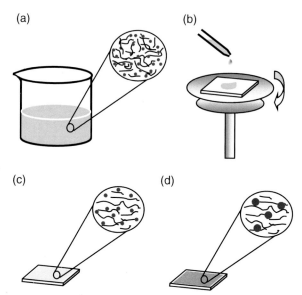

Figure 6.1. *In situ* fabrication of polymer–metal nanocomposite thin film: **(a)** solution of the polymer and metal precursor, **(b)** spin coating of the solution on a suitable substrate, **(c)** thin film of the polymer and metal precursor on the substrate, and **(d)** thin film of the polymer with embedded metal nanoparticles on the substrate.

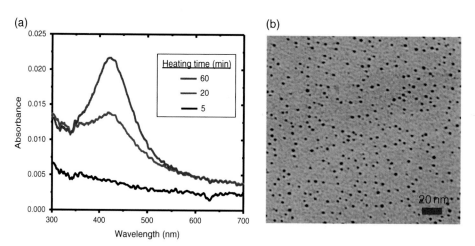

Figure 6.3. **(a)** Optical absorption spectrum of Ag–PVA film (Ag/PVA weight ratio = 0.028) fabricated by heating at 90 °C for different periods of time. **(b)** TEM image of Ag–PVA film (Ag/PVA weight ratio = 0.028) fabricated by heating at 70 °C for 60 min. Figures adapted with permission from Reference [8]. Copyright © 2005, American Chemical Society.

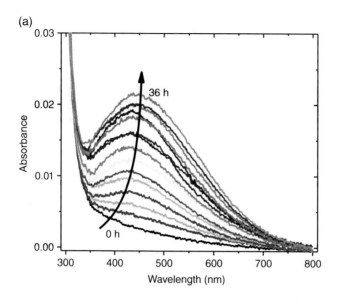

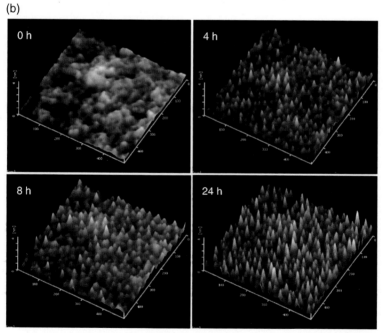

Figure 6.4. Real-time monitoring of the growth of silver nanoparticles at 25 °C in a PVP thin film: **(a)** the optical absorption spectra and **(b)** AFM images recorded at different times. Figures adapted with permission from Reference [13]. Copyright © 2009, Royal Society of Chemistry.

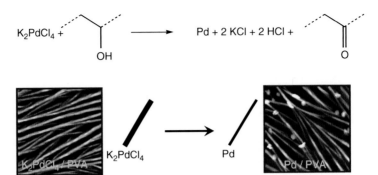

Figure 6.6. Chemical reaction occurring inside the polymer thin film in the reduction of K_2PdCl_4 to Pd by PVA. AFM images showing the crystal-to-crystal transformation of K_2PdCl_4 to Pd nanowires; the particles in the image of the product are KCl microcrystals formed as by-products in the reaction. Figure adapted with permission from Reference [15]. Copyright © 2007, John Wiley & Sons, Inc.

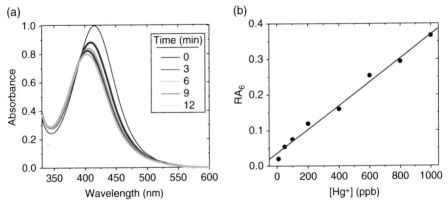

Figure 6.8. (a) Temporal variation of the LSPR spectra of Ag–PVA thin film immersed in a 400 ppb aqueous solution of Hg^+; absorbance at λ_{max} at zero time is normalized to 1.0. (b) Relative change in the absorbance at λ_{max} of the Ag–PVA film in 6 min (RA_6), as a function of the concentration of Hg^+; the least square fit to a straight line is indicated. Figures adapted with permission from Reference [31]. Copyright © 2011, American Chemical Society.

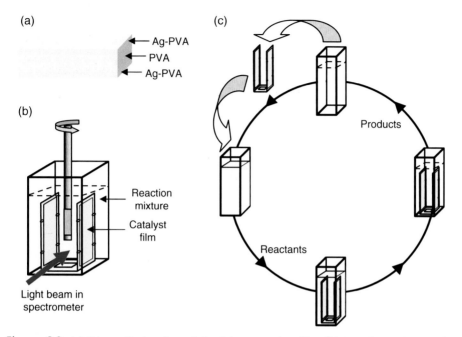

(a)

Ag-PVA
PVA
Ag-PVA

(b)

Reaction
mixture

Catalyst
film

Light beam in
spectrometer

(c)

Products

Reactants

Figure 6.9. (a) Schematic drawing of the 3-layer catalyst film. (b) Experimental setup for monitoring the reduction of 4-NP by $NaBH_4$, catalyzed by the Ag-PVA catalyst film. Figure adapted with permission from Reference [12]. Copyright © 2007, John Wiley & Sons, Inc. (c) Schematic representation of the "dip catalyst" reuse cycle.

(a)

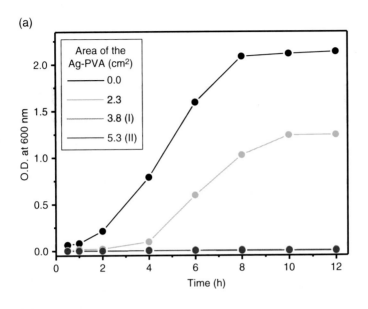

(b)

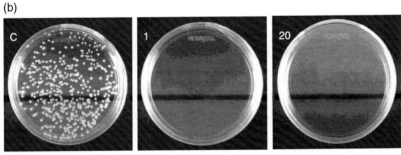

Figure 6.10. (a) Bacterial growth reflected in the increase in optical density (at 600 nm) of *E. coli* suspension with time; control (no film) and samples treated with Ag-PVA film having different areas and hence silver content are shown; I and II correspond to the minimum inhibitory and minimum bactericidal concentrations. (b) Photographs of petri plates (after 12 h incubation) spread with water samples inoculated with 10^5 CFU *E. coli*, control (C), and those treated for 15 min with the same Ag-PVA film (First and twentieth use) are shown. Figure adapted with permission from Reference [35]. Copyright © 2011, Indian Academy of Sciences.

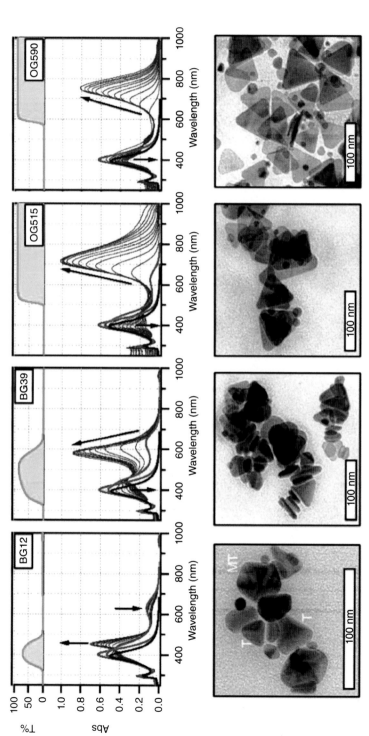

Figure 7.4. Sample evolution and results of photochemical silver NP preparation under different illumination conditions. (Top) Evolution monitored by UV–Vis absorption spectra taken at 30 min intervals: black, NP solution before irradiation; blue, induction period of exposition by using a conventional fluorescent tube; red, second stage of growth by sample illumination fitted with a specific colored filter. The transmission of the filters used is indicated by the green shaded area on top of each graph. The arrows indicate the evolution of the main spectral features during this stage. (Bottom) TEM micrographs of the resulting silver NPs. Reprinted with permission from Reference [53]. Copyright (2003) American Chemical Society.

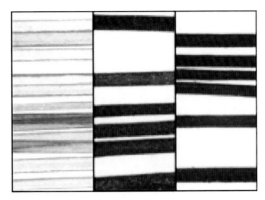

Figure 8.1. Light photos show blonde hair becoming increasingly darker after multiple applications of hair dye recreated from the ancient Greco-Roman recipe. Image adapted with permission from [36]. Copyright (2006) American Chemical Society.

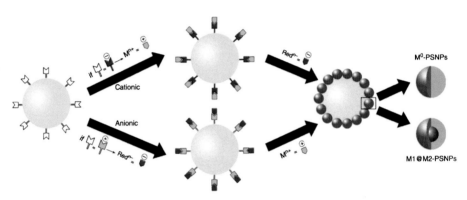

Figure 8.3. Schematic diagram of IMS steps for synthesis of NPs in granulated polymer matrix.

(a)   (b)

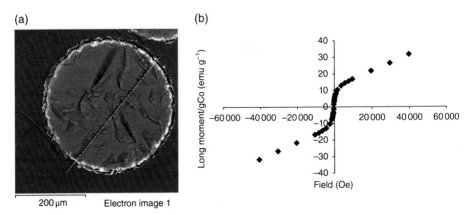

Figure 8.4. (a) Scanning electron microscope image of cross section of granulated Co MNP–sulfonated polymer NC; red line across bead diameter corresponds to EDS spectrum, which shows Co MNP distribution and (b) magnetization curve of NC sample.

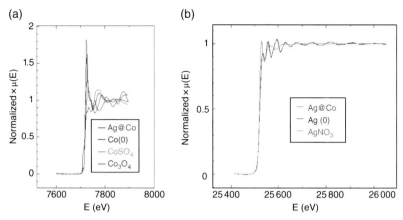

Figure 8.7. X-ray absorption near-edge structure spectra of Co (a) and Ag (b) signals of the metal Ag@Co carboxylated resin sample (blue line in both graphics) in comparison with Co and Ag standards. (a) Co^0 (red), $CoSO_4 \cdot 7H_2O$ (green), Co_3O_4 (violet), and (b) Ag^0 (red) and $AgNO_3$ (green) are standards.

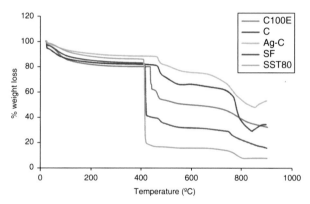

Figure 8.8. Thermogravimetric analysis curves of weight loss versus temperature for sulfonated polymer samples: C100E NP-free polymer; C sample (Fe_3O_4–C100E); Ag-C sample (Ag@Fe_3O_4–C100E); SST80 NP-free polymer and SF sample (Fe_3O_4–SST80) (see also Table 8.2).

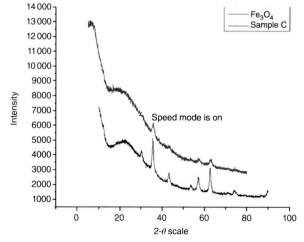

Figure 8.9. X-ray diffraction patterns of magnetite NPs: polymer-free (black line) and in sample C (red line).

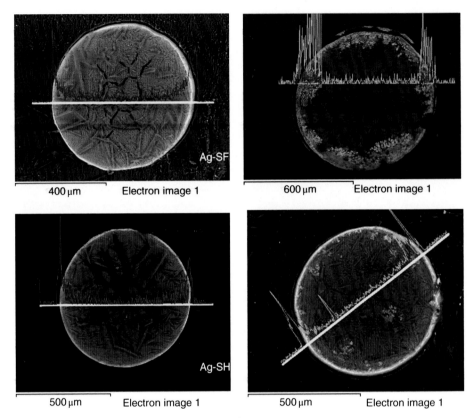

Figure 8.10. Scanning electron microscopy images of the Fe_3O_4 and Ag@Fe_3O_4 NC cross sections for samples SF, Ag-SF, SH, and Ag-SH (see Table 8.1). The lines show distribution of Ag (blue) and Fe (red) metal ions across particle diameters.

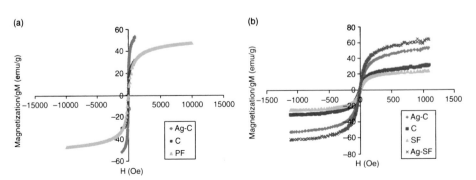

Figure 8.12. Magnetization curves (VSM graphs) for Fe_3O_4 and Ag@Fe_3O_4 NPs in both sulfonated C100E and SST80 polymers. (a) Comparison between Fe_3O_4 and Ag@Fe_3O_4 NPs in C100E polymer; (b) comparison between Fe_3O_4 and Ag@Fe_3O_4 NPs in C100E and SST80 polymers.

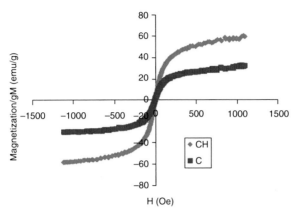

Figure 8.13. Magnetization curves (VSM graphs) for Fe_3O_4 and $Ag@Fe_3O_4$ NP magnetite prepared using lower concentrations of both metals (Fe and Ag) in sulfonic polymer, C100E (CH sample compared to C sample).

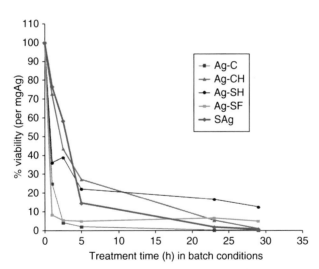

Figure 8.14. Cell viability versus treatment time for $Ag@Fe_3O_4$ NP NCs in both C100E and SST80 polymers and for monometallic Ag NP–SST80 NCs.

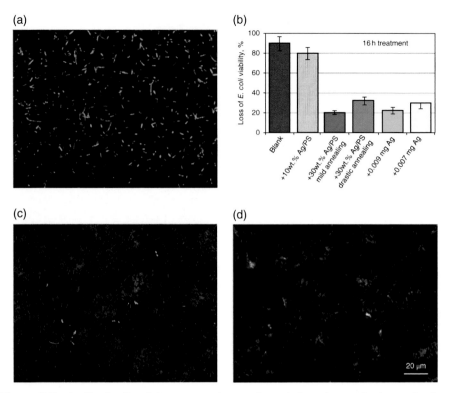

Figure 9.7. Live/dead cell staining assay: microscopic analysis and quantitative evaluation. (a) High number of green-labeled living bacteria after 16h incubation in LB, without Ag/PS. (c, d) High number of red-labeled dead bacteria after 16h incubation +30wt.% Ag/PS film (1 cm × 1 cm – 0.0177 mg) obtained by using a mild thermal annealing treatment (C); similar results were observed in bacteria culture after 16 h of incubation with Ag powder (D). Scale bar represents 20 μm. (b) Quantitative analysis of loss of *E. coli* viability (%) after 16 h incubation at each condition reported in Table 9.1 (A–E).

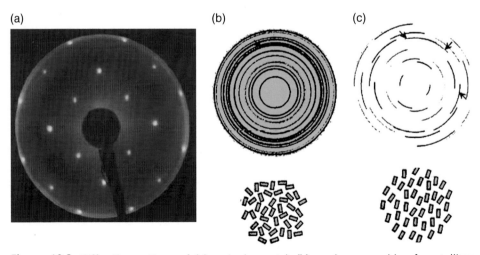

Figure 10.2. Diffraction pattern of (a) a single crystal, (b) random assembly of crystallites (polycrystal), (c) polycrystal with preferred orientations. Reprinted from http://www.crystal.mat. ethz.ch/research/ZeolitesPowderDiffraction/Texture.

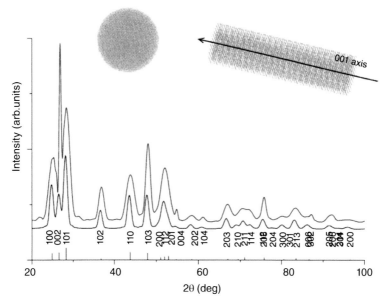

Figure 10.4. Diffraction profiles of CdSe wurtzite nanocrystals with different shapes: sphere (black profile) and rod (red profile).

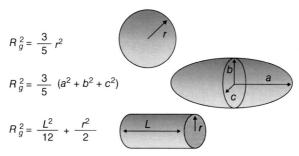

$$R_g^2 = \frac{3}{5}\, r^2$$

$$R_g^2 = \frac{3}{5}\, (a^2 + b^2 + c^2)$$

$$R_g^2 = \frac{L^2}{12} + \frac{r^2}{2}$$

Figure 10.7. Radius of gyration (R_g) for different shapes.

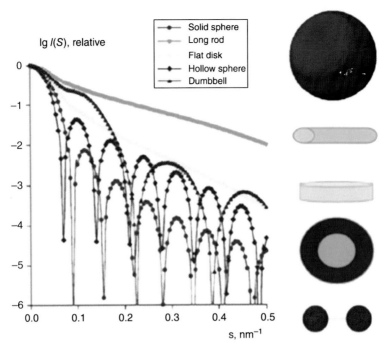

Figure 10.8. SAXS patterns for different shapes. Reprinted with permission from Reference [20].

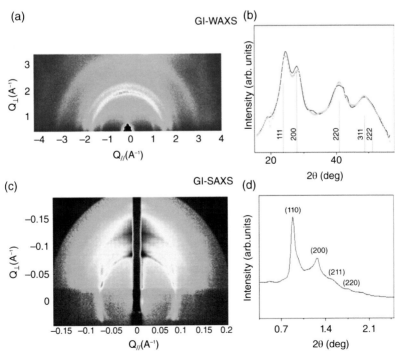

Figure 10.9. GIWAXS (**a, b**) and GISAXS (**c, d**) data collected on a 3D assembly of PbS nanocrystals. Reprinted with permission from Reference [25].

6

CHEMISTRY INSIDE A POLYMER THIN FILM: *IN SITU* SOFT CHEMICAL SYNTHESIS OF METAL NANOPARTICLES AND APPLICATIONS

E. Hariprasad and T. P. Radhakrishnan

School of Chemistry, University of Hyderabad, Hyderabad, India

6.1 INTRODUCTION

Metals have played a central role in the evolution and progress of human civilization across centuries. It is only natural that nanometric metal particles with their unique physical and chemical attributes have fascinated scientists all through. Whether it be the practical utilization of *Swarna Bhasma* in traditional Indian Ayurvedic medicine [1] or the fundamental exploration of the optical properties of *finely divided gold* by Michael Faraday [2], nanometals have made a mark in science and technology from early on. Recent years have witnessed an explosive growth in the field of metal nanoparticles, development of fabrication methodologies, investigation of the unique characteristics, and deployment in a wide range of applications from electronics and photonics to catalysis and sensing to biology and medicine [3–5].

Metal nanoparticles are fabricated, investigated, and utilized in various formulations such as aqueous or organic solutions, composites, powders, organized assemblies, and thin films. Even though metals and polymers in general exhibit contrasting

Nanocomposites: In Situ *Synthesis of Polymer-Embedded Nanostructures*, First Edition.
Edited by Luigi Nicolais and Gianfranco Carotenuto.

characteristics in their bulk states, formation of a polymer–metal nanocomposite provides a versatile route to harmonize the two and harness unique benefits. The metal nanoparticles and the polymer matrix can impact each other in mutualistic ways, leading to the emergence of novel characteristics and functions. Thin films of these nanocomposite materials are particularly advantageous in terms of the ease of fabrication as well as the breadth of applications. Even though a wide range of protocols are available for the preparation of polymer–metal nanocomposite thin films, the various *in situ* routes that have been developed are especially attractive due to their simplicity, adaptability, and wide applicability [6]. Generally, these techniques involve the casting of thin polymer films with the metal nanoparticle precursors already embedded within or introduced from the external medium, followed by a wide range of physical and chemical processes through which the nanoparticles are generated in the polymer matrix.

In this chapter we describe the versatility of the *in situ* approach developed in our laboratory over the past few years, for the generation of noble metal nanoparticles within a polymer thin film using soft chemistry carried out within the film. Highlights of this method include the utilization of the polymer itself as the reducing agent and stabilizer for the formation of metal nanoparticles, mild thermal annealing used for the precursor-to-product transformation, feasibility of fabricating free-standing nanocomposite thin films, and the environmentally benign conditions of synthesis. An important aspect that will be discussed is the facile monitoring of the chemistry within the polymer film and the nanoparticle formation, through a variety of spectroscopy and microscopy tools. Unique and superior advantages of employing the nanoparticle-embedded polymer thin film in sensing, catalysis, and antibacterial applications, exploiting the chemical processes within the film, will be described. Other advanced materials application of these organic–inorganic hybrid materials will be noted briefly.

6.2 *IN SITU* FABRICATION AND MONITORING

Chemical reactions in the gas and solid phases are not uncommon; however, the majority of chemical synthesis is still carried out in the liquid (solution) phase. Reactions inside a solid polymer matrix provide an elegant alternative to synthesize polymer–metal composites. The essential requisite to achieve a soft chemical synthesis is the appropriate choice of polymer possessing functional groups that can act as the reducing agent for the metal ion of interest. A range of reaction conditions including mild thermal treatment can be used to induce the metal ion to metal atom transformation with concomitant oxidation of the polymer functionality. As the diffusion of ions and atoms inside the macromolecular maze is relatively restricted, this approach is ideally suited to control the nucleation and growth of the product, leading to the generation of nanostructures. The popular polyol protocol suggests that a convenient choice of polymer for such a synthesis of noble metal nanoparticles is poly(vinyl alcohol) (PVA); overpotential effects may be relevant as pointed out in the case of polyol reduction [7]. Indeed, a range of other polymers with functionalities such as amine and alcohol groups can be chosen as well.

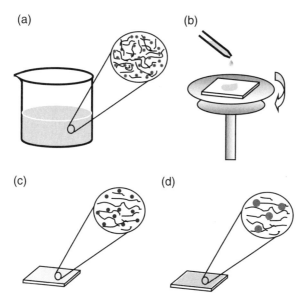

Figure 6.1. *In situ* fabrication of polymer–metal nanocomposite thin film: **(a)** solution of the polymer and metal precursor, **(b)** spin coating of the solution on a suitable substrate, **(c)** thin film of the polymer and metal precursor on the substrate, and **(d)** thin film of the polymer with embedded metal nanoparticles on the substrate. (*See insert for color representation of the figure.*)

The protocol for the *in situ* synthesis of metal nanoparticles through chemistry carried out inside the polymer thin film is illustrated schematically in Figure 6.1. A typical procedure begins with the preparation of an aqueous solution of the polymer and the metal ion precursor. These solutions are mixed in desired proportions and a few drops spin coated on a suitable substrate like glass or quartz. The thin film is subsequently heated at a chosen temperature for a specific time whereupon the metal ion is reduced by the functional groups on the polymer to the metal atom, and the atoms aggregate to form the nanoparticles. The chemistry happening inside the polymer can be described using the typical reaction sequences shown in Figure 6.2. The reduction process may be followed by significant chemical changes in the polymer such as the cross-linking shown in Figure 6.2.

A useful variant of the procedure outlined above involves the pre-coating of the substrate with a sacrificial polymer like polystyrene upon which the solution of the reducing polymer and metal precursor is spin coated. Following the thermal treatment and generation of the nanoparticles, the polystyrene sub-layer can be dissolved in an organic solvent to release freestanding films of the polymer–metal nanocomposite thin film. By suitable choice of solution dilution and spinning conditions, it is possible to make the final nanocomposite film thin enough for direct imaging in a transmission electron microscope (TEM). This is a critical advantage for the characterization of the metal nanoparticles embedded within the polymer film.

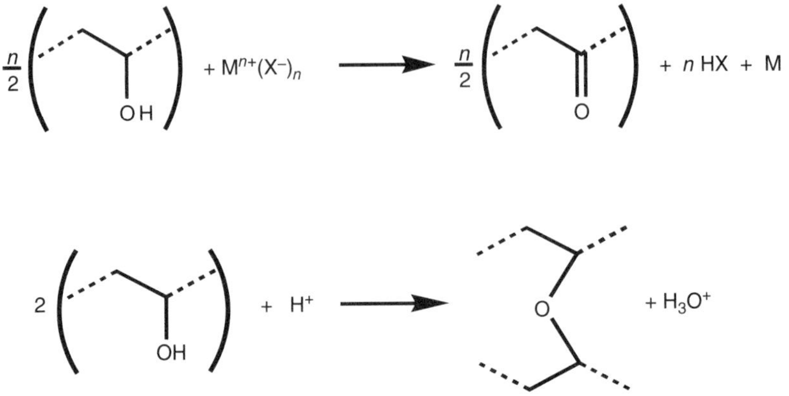

$$\frac{n}{2}\left(\cdots\!\!\!\bigwedge_{\text{OH}}\!\!\!\cdots\right) + M^{n+}(X^-)_n \longrightarrow \frac{n}{2}\left(\cdots\!\!\!\bigwedge_{\text{O}}\!\!\!\cdots\right) + n\,HX + M$$

$$2\left(\cdots\!\!\!\bigwedge_{\text{OH}}\!\!\!\cdots\right) + H^+ \longrightarrow \Big(\cdots O \cdots\Big) + H_3O^+$$

Figure 6.2. Reaction schemes showing the reduction of metal ion by PVA and a possible polymer cross-linking pathway.

Besides the ease and simplicity of carrying out the *in situ* synthesis of metal nanoparticles, an attractive feature of the chemistry inside the polymer film is the facility with which it can be monitored using a wide range of tools. Immobilization of the nanoparticles by the polymer matrix is the key factor here. The thin film nature makes it particularly convenient to record spectroscopic or diffraction data as well as microscopy images through the course of the reaction; this could be done on the thin films in real time or on samples retrieved by quenching the reaction at various stages. The specific systems discussed in the following sections demonstrate the versatility of the *in situ* synthesis as well as the ease of monitoring the chemistry and the nanoparticle growth.

6.2.1 Silver

PVA with a range of average molecular weights and extents of hydrolysis have been found suitable to prepare silver nanoparticles from silver salts, through the *in situ* protocol described above. Aqueous solutions of silver nitrate ($AgNO_3$) and PVA were mixed and spin coated. A detailed study showed that optimization of the Ag/PVA ratio and the time and temperature of the thermal treatment can yield Ag–PVA thin films containing highly monodisperse silver nanoparticles, typically a few nanometers in size [8]. Formation of the nanoparticles in the film is demonstrated by the emergence and growth of the localized surface plasmon resonance (LSPR) absorption (extinction); optical absorption spectra of the unheated film and films heated for different periods of time are shown in Figure 6.3a. Direct imaging of the film in a TEM using the methodology described above reveals nanoparticles in the size range 2–3 nm (Fig. 6.3b). X-ray diffraction has been used to monitor the changes occurring in the film during the *in situ* growth of the nanoparticles [9]. Spectroscopic ellipsometry is a powerful tool to assess the formation of the nanoparticles within the film [10]. Spin coating multiple layers with thermal treatments in between has been used to fabricate different kinds of multilayer structures such as PVA/Ag–PVA/PVA [11] and Ag–PVA/PVA/Ag–PVA [12] of special interest in particular applications described later.

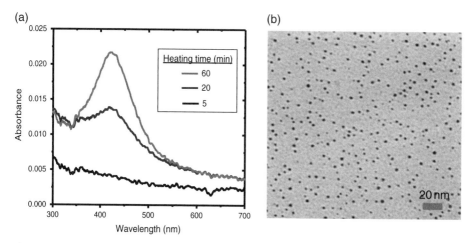

Figure 6.3. (a) Optical absorption spectrum of Ag–PVA film (Ag/PVA weight ratio = 0.028) fabricated by heating at 90 °C for different periods of time. (b) TEM image of Ag–PVA film (Ag/PVA weight ratio = 0.028) fabricated by heating at 70 °C for 60 min. Figures adapted with permission from Reference [8]. Copyright © 2005, American Chemical Society. (*See insert for color representation of the figure.*)

An important advantage of the *in situ* fabrication protocol is the feasibility of monitoring the formation of metal nanoparticles in real time. If the process were to occur under ambient conditions, this would be a particularly simple exercise. It was observed that spin-coated thin films of poly(vinyl pyrrolidone) (PVP) containing $AgNO_3$ shows spontaneous formation of silver nanoparticles at 25 °C under ambient atmosphere over periods extending up to 72 h [13]. The growth is reflected in the steady evolution of the LSPR absorption (Fig. 6.4a). Growth of the nanostructures can also be recorded in an atomic force microscope (AFM); a typical series of topography images are shown in Figure 6.4b. TEM studies suggested that the observed structures are likely to be aggregates of silver nanoparticles embedded within the polymer. TEM images of the film confirm the formation of silver nanoparticles under the ambient conditions.

6.2.2 Gold

The high reduction potential of Au^{3+} makes it particularly easy to generate gold atoms by reduction, which aggregate to form gold nanoparticles. *In situ* reduction of $HAuCl_4$ inside PVA thin film is found to be very facile, leading to the formation of gold nanoplates of various polygonal shapes [14]. Careful control of the Au/PVA weight ratio and the thermal annealing conditions results in the formation of hexagonal, pentagonal, rectangular, and triangular nanoplates; interestingly, the plates form with a preferential orientation of the (111) face parallel to the substrate plane. Figure 6.5 shows the TEM image of triangular and pentagonal gold nanoplates formed *in situ* inside the PVA film. An infrared spectroscopic study demonstrated the oxidation of the hydroxy groups on PVA accompanying the reduction of Au^{3+} during the fabrication of Au–PVA thin films.

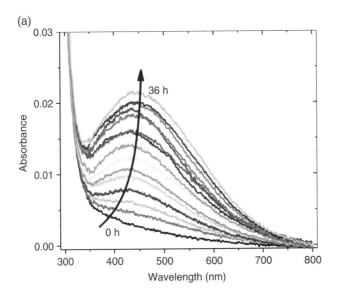

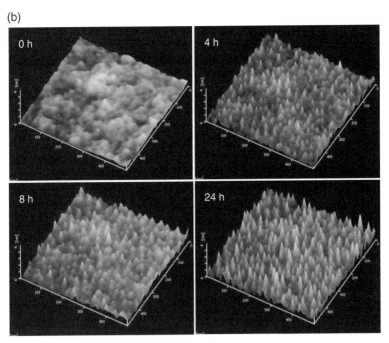

Figure 6.4. Real-time monitoring of the growth of silver nanoparticles at 25 °C in a PVP thin film: **(a)** the optical absorption spectra and **(b)** AFM images recorded at different times. Figures adapted with permission from Reference [13]. Copyright © 2009, Royal Society of Chemistry. (*See insert for color representation of the figure.*)

Figure 6.5. TEM image of Au–PVA film (Au/PVA weight ratio = 0.12) fabricated by heating at 130 °C for 60 min. Figure taken with permission from Reference [14]. Copyright © 2005, Royal Society of Chemistry.

6.2.3 Palladium

K_2PdCl_4 served as an ideal precursor for the *in situ* synthesis of nanopalladium inside the PVA matrix [15]. Interestingly, when the concentrations were sufficiently high, the precursor salt crystallized inside the polymer thin film in the form of dendritic nanowires. Thermal annealing, typically at 130 °C for 2 h, led to a transformation of the precursor nanowires to Pd nanowires. This could be established through X-ray photoelectron spectroscopy, scanning electron microscopy, AFM, and TEM together with selected area electron diffraction (SAED). The experiments revealed that the reduction of Pd^{2+} is generally incomplete and that the final nanocomposite thin films contained the precursor and product nanowires. Most importantly, the microscopy studies showed clearly the formation of microcrystals of the by-product KCl within the polymer matrix. The *in situ* synthesis of Pd nanowires thus represents an elegant example of soft chemistry carried out within a polymer thin film, resulting in a nanocrystal-to-nanocrystal transformation (Fig. 6.6).

6.2.4 Platinum

A brief report is available on the *in situ* generation of platinum nanoparticles inside PVA film [16]. Thick films of PVA containing H_2PtCl_6 were cast in petri dishes, dried over several days, and then subjected to thermal annealing. Formation of platinum nanoparticles, typically tens of nanometers in size, was observed.

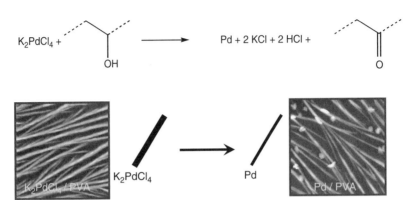

Figure 6.6. Chemical reaction occurring inside the polymer thin film in the reduction of K_2PdCl_4 to Pd by PVA. AFM images showing the crystal-to-crystal transformation of K_2PdCl_4 to Pd nanowires; the particles in the image of the product are KCl microcrystals formed as by-products in the reaction. Figure adapted with permission from Reference [15]. Copyright © 2007, John Wiley & Sons, Inc. (*See insert for color representation of the figure.*)

6.2.5 Mercury

The unique status of mercury among elements, and metals in particular, makes the fabrication of its nanostructures fundamentally important and, at the same time, challenging. Even though a few studies on nanomercury trapped in porous glass [17, 18] or in the form of colloidal solution [19, 20] are known, directed synthesis of mercury nanodrops and stable nanocrystals and direct observation and characterization of their unique and size-dependent attributes have been reported only recently [21]. The physical state of mercury under ambient temperature conditions rules out the possibility of deploying most of the common approaches to nanometal fabrication. Exploitation of the soft chemical reduction of mercury ions *in situ* inside a polymer such as PVA provides a singularly facile and efficient route to the generation of nanomercury embedded within the polymer film. Spin-coated films of PVA containing $Hg_2(NO_3)_2$ when heated at 110 °C for 1 h exhibited an absorption with λ_{max} at 287 nm (Fig. 6.7a) [21] notably, this is in good agreement with the LSPR absorption predicted for nanomercury [22]. Imaging in a TEM at the ambient temperature of 23 °C showed that the mercury nanodrops evaporate under the electron beam. However, when cooled to −120 °C, the nanodrops freeze to form stable nanocrystals (Fig. 6.7b) as shown by the electron diffraction (Fig. 6.7c) as well as the high-resolution image (Fig. 6.7d), consistent with the known rhombohedral structure of crystalline mercury [23]. Most interestingly, the nanocrystals can be melted within the TEM by increasing the temperature and frozen back by cooling again. These experiments revealed strong hysteresis effects in the melt–freeze cycles. The melting temperature is found to decrease with the size of the mercury nanocrystal, a phenomenon well known in nanoparticles [24–26]; the depression with respect to the bulk value is found to be inversely proportional to the diameter as expected based on the Gibbs–Thomson equation.

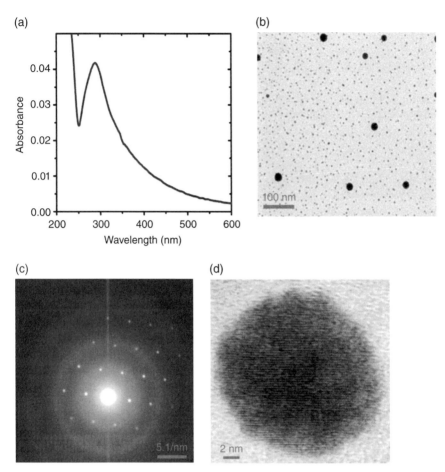

Figure 6.7. (a) Optical absorption spectrum and (b) TEM image (recorded at –120 °C) of Hg–PVA film. (c) Electron diffraction from and (d) high-resolution image of a mercury nano-crystal in the Hg–PVA film (recorded at –120 °C). Figures adapted with permission from Reference [21]. Copyright © 2011, American Chemical Society.

6.3 APPLICATIONS

The various examples of fabrication of metal nanoparticle-embedded polymer thin films described above highlight the simplicity of the methodology used. The environmentally benign nature of the protocol mostly involving aqueous medium, the facile spin-coating process used to form thin films, and the mild conditions required for the soft chemical synthesis of the nanoparticles are important features. The procedure ensures that the potentially hazardous metal nanoparticles are formed within the polymer film and therefore never exposed. It is notable that the procedure is highly scalable and can be easily adapted for coating surfaces with large area as well as irregular shapes. The feasibility of fabricating free-standing as well as multilayer films is relevant not only in basic characterization procedures as noted earlier but also in several applications. All these factors indicate that the methodology we have developed for the *in situ* generation of metal nanoparticles inside polymer thin films is cost effective and efficient and hence of potential interest in a wide range of applications that metal nanoparticles are employed in.

The metal nanoparticle-embedded polymer thin films formed through the *in situ* approach possess unique attributes that effectively augment their application potential. While the aqueous solubility of the precursor and the polymer are advantageous in terms of making the film formation facile and environmentally benign as noted earlier, the likelihood of the polymer chain cross-linking simultaneously with the nanoparticle production diminishes the solubility of the final nanocomposite thin film, enabling its application in aqueous media. At the same time, the hydrogel character of polymers like PVA is critically important, as it leads to the swelling of the thin film, facilitating contact between the external medium and the metal nanoparticles embedded within the film. Careful choice of the polymer matrix can ensure that the nanoparticles do not leach out of the film during such interactions. This in turn opens up the opportunity to reuse the nanoparticle-embedded polymer thin film in multiple cycles, a factor critically important in a variety of applications as discussed in the following sections. Another major feature of fundamental interest is the feasibility of monitoring the nanoparticles through the usage cycles by a variety of analytical techniques. This can be achieved through direct examination of the whole film *in situ* or analysis of small portions of the film removed periodically. We describe below, some of the applications explored mainly using Ag–PVA thin films fabricated using the *in situ* method.

6.3.1 Sensor

The LSPR absorption of metal nanoparticles is an effective tool commonly used in sensor applications. Even though there is extensive activity in the field of mercury sensing, majority of the studies have focused on the detection of Hg^{2+} ions, and many of the sensors work only in the *in situ* mode [27–30]. Ag–PVA thin film has been shown to be a highly efficient, fast, and selective sensor for mercury in all its stable oxidation states, (0, +1, +2), in aqueous medium, and in both *in situ* and *ex situ* modes [31]. Sensitive changes in the LSPR absorption of the Ag–PVA film in terms of both reduction of the intensity and a characteristic blue shift of the λ_{max} (Fig. 6.8a) have been exploited for detecting mercury down to ppb levels. While the mercury ions enter into galvanic replacement reaction with the silver nanoparticles, neutral mercury causes amalgamation. The chemistry within the polymer film causes clear and reproducible changes in the absorbance of the Ag–PVA film within minutes of exposure; the relative absorbance change is found to increase linearly with mercury concentration in the ppb–ppm range (Fig. 6.8b).

While the LSPR absorption of Ag–PVA is insensitive to most of the transition and alkali metal ions, the distinct blue shift upon contact with mercury is in sharp contrast with the red shift occurring with ions such as Au^{3+}, making the mercury sensing highly selective. As the modification of the silver nanoparticles embedded within the polymer

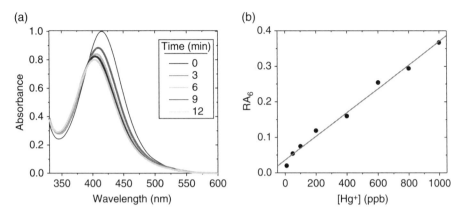

Figure 6.8. (a) Temporal variation of the LSPR spectra of Ag–PVA thin film immersed in a 400 ppb aqueous solution of Hg⁺; absorbance at λ_{max} at zero time is normalized to 1.0. (b) Relative change in the absorbance at λ_{max} of the Ag–PVA film in 6 min (RA₆), as a function of the concentration of Hg⁺; the least square fit to a straight line is indicated. Figures adapted with permission from Reference [31]. Copyright © 2011, American Chemical Society. (*See insert for color representation of the figure.*)

film upon contact with mercury and its ions in water is permanently registered in the Ag–PVA film, its LSPR absorption can be examined at a later time, enabling *ex situ* sensing and monitoring. This is a major advantage of the thin film sensor in practical applications. As stated earlier, the thin film nature of the film also allowed extensive and detailed spectroscopic and microscopic examination of the film after the sensing process, so that the basic transformations of the silver nanoparticles that occurred within the polymer matrix could be unraveled.

6.3.2 Catalyst

The large surface-to-volume ratio and in some cases the specific surface structure of nanoparticles form the basis of their efficient use as catalysts in chemical reactions. Even though nanoparticles are expected to effectively harmonize the advantages of homogeneous and heterogeneous catalysts [32, 33], difficulties with their recovery and reuse continue to be a major challenge. If they can be incorporated inside a suitable matrix and still be made accessible to the reactants, a highly reusable and efficient catalyst can be realized. This has been demonstrated using Ag–PVA film as the catalyst in the prototypical reaction involving the reduction of 4-nitrophenol (4-NP) by sodium borohydride ($NaBH_4$) [12]. An important factor in this experiment was the design of the thin film catalyst; based on a number of trial experiments aimed at maximizing the efficiency and reusability, a 3-layer structure Ag–PVA/PVA/Ag–PVA (Fig. 6.9a) was developed, which ensured sufficient robustness for the film and high accessibility of the silver nanoparticles.

The reaction kinetics could be easily monitored by optical absorption spectroscopy with no interference from the catalyst as the film could be kept out of the light beam

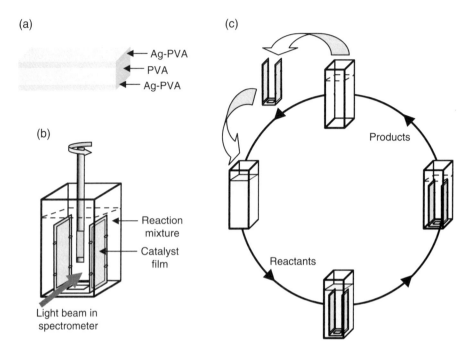

Figure 6.9. (a) Schematic drawing of the 3-layer catalyst film. (b) Experimental setup for monitoring the reduction of 4-NP by NaBH$_4$, catalyzed by the Ag-PVA catalyst film. Figure adapted with permission from Reference [12]. Copyright © 2007, John Wiley & Sons, Inc. (c) Schematic representation of the "dip catalyst" reuse cycle. (*See insert for color representation of the figure.*)

path (Fig. 6.9b). An important observation is that the reaction can be instantaneously switched on or off by dipping the nanocomposite thin film catalyst in or taking it out; this rules out the possibility of the catalysis being induced by silver nanoparticles leached out of the film. It also forms the basis for describing the nanoparticle-embedded polymer thin film as a "dip catalyst" [12]. A further important implication is the feasibility of convenient multiple uses of the same catalyst film. Figure 6.9c is a schematic representation of the cycle; introduction of the catalyst film induces the reaction as shown by the disappearance of the yellow color of 4-nitrophenolate, following which the film is withdrawn, washed, dried, and reintroduced in a new reaction batch. It was found that even after 30 reuses, the catalytic activity is only marginally compromised, the reaction yield coming down from ~99% to ~96%. The thin film catalyst could be periodically examined using the LSPR absorption spectrum, TEM, and AFM; while the film was largely unperturbed, slight particle aggregation leading to some changes in the size distribution could be observed after several reuses.

The *in situ* protocol with PVA allows the production of nanocomposite thin films containing nanoparticles of the various well-known noble metal catalysts. Utility of such "dip catalysts" in chemical transformations remains to be explored. By definition, a catalyst does not undergo any change during its intervention in a reaction and hence should be capable of endless repeated uses. The "dip catalyst" is a major step towards realizing this ideal.

6.3.3 Antibacterial

Silver nanoparticles are well-known antibacterial agents [34]. However, an effective formulation that ensures facile synthesis and fabrication, large area application, high efficacy, and multiple reuse capability is yet to be developed. An exploratory study on *Escherichia coli* showed that Ag–PVA film can act as a potent and reusable bactericide [35]. The minimum inhibitory and bactericidal concentrations (of silver in the nano-composite film used to treat unit volume of the *E. coli* bacterial medium) are found to be 4.5 and 6.5 µg ml^{-1}, respectively, indicating the high efficacy of the Ag–PVA film (Fig. 6.10a). It was also found that one piece of film can be used to repeatedly and efficiently kill the bacteria in several samples of water, with simple washing in between the usages (Fig. 6.10b). Examination of the film between uses by AFM demonstrated that it remained intact throughout. The small reduction in the LSPR absorption over a large number of reuses revealed the minor extent of leaching of the nanoparticles. Effective bactericidal activity of Ag–PVA film coated on a glass rod was demonstrated by the significant reduction of bacterial colonies in water samples stirred with the rod for a few minutes. This experiment illustrates the promise of the *in situ* synthesized nanoparticle-embedded polymer thin film in important applications such as water purification.

6.3.4 Miscellaneous

Several applications of the *in situ* synthesized metal nanoparticle-embedded polymer thin films that involve physical responses of the nanocomposite thin films have also been explored. As metal nanoparticles in colloidal medium are known to show strong optical limiting responses, the thin films have been investigated in this context. Ag–PVA films exhibit appreciable optical limiting characteristics for ns- and fs-pulsed lasers [8, 11, 36]; a notable point is the high transparency in the linear regime. Studies on free-standing PVA/Ag–PVA/PVA films allowed the estimation of accurate nonlinear coefficients without any error due to contributions from the substrate [11]. Optical limiting capability of Au:Ag–PVA [37], Pd–PVA [15], and Pt–PVA [16] thin films has also been explored. An interesting discovery with great application potential is the random lasing with coherent feedback observed in Ag–PVA thin films containing a dye molecule like Rhodamine-6G [38]. *In situ* synthesized Ag–PVA film containing poly(γ-glutamic acid) has been shown to be an efficient substrate for surface-enhanced Raman scattering [39].

Ag–PVA films show appreciable microwave absorption in the frequency range of 8–12 GHz of interest in telecommunication applications [40]. *In situ* generation of silver and gold nanoparticles inside a PVA thin film with concomitant polymer cross-linking induced by electron beam lithography allows pattern formation [41, 42]. This allows the generation of negative-tone polymer nanocomposite resists, which are of interest in plasmonic technologies. Nanocomposite thin films fabricated by the *in situ* technique have also been shown to exhibit significant dichroic properties upon uniaxial drawing [43] and to act as efficient membranes for selective gas transport [44].

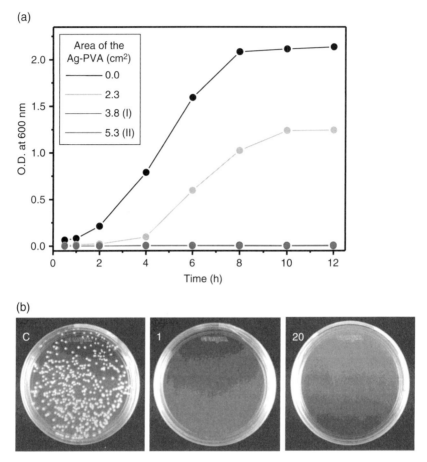

Figure 6.10. (a) Bacterial growth reflected in the increase in optical density (at 600 nm) of *E. coli* suspension with time; control (no film) and samples treated with Ag-PVA film having different areas and hence silver content are shown; I and II correspond to the minimum inhibitory and minimum bactericidal concentrations. **(b)** Photographs of petri plates (after 12 h incubation) spread with water samples inoculated with 10^5 CFU *E. coli*, control (C), and those treated for 15 min with the same Ag-PVA film (First and twentieth use) are shown. Figure adapted with permission from Reference [35]. Copyright © 2011, Indian Academy of Sciences. (*See insert for color representation of the figure.*)

6.4 CONCLUSIONS

Chemistry within the polymer thin film offers a unique and elegant route to synthesize metal nanoparticles in a ready-to-use form. Specific choice of the metal precursor and polymer ensures that no additional reducing agent is required, mild conditions are sufficient for the soft chemical process, and the nanoparticles generated *in situ* are stabilized by the polymer matrix itself. The various studies discussed in this chapter demonstrate that the parameters involved in the fabrication, such as the metal/polymer ratio as well as the thin film coating and thermal treatment conditions, provide fine control on the size, shape, and distribution of the nanostructures generated. Convenient monitoring of the fabrication of the nanocomposite thin film by various spectroscopy and microscopy tools is a significant advantage of the protocol described. The possibility of carrying out chemistry within the polymer film takes on a new dimension in applications such as sensing and catalysis. The *in situ* synthesized polymer–metal nanocomposite thin film can also function as a potent bactericide, an optical limiter, and a microwave absorber.

An important aspect of the nanocomposite thin films described is their amenability to scale up and deployment in practical device applications. This is an area that is likely to be explored in greater detail in the near future. Another important aspect that is beginning to be probed is the control of nanoparticle assembly within the polymer film, exploiting synergistic effects between the metal and the polymer. This will have significant implications for the design and development of electronics and photonics devices based on these nanocomposite materials.

ACKNOWLEDGMENTS

Significant contributions to the development of the work described in this chapter came from Dr. Shatabdi Porel, Dr. G. V. Ramesh, and Mr. Muvva D. Prasad. In addition to them, we thank also our collaborators, Prof. D. Narayana Rao, Prof. James Raju and Prof. Aparna Dutta Gupta (University of Hyderabad); Dr. Shashi Singh (Centre for Cellular and Molecular Biology, Hyderabad); and Dr. B. Sreedhar (Indian Institute of Chemical Technology, Hyderabad). Financial support from the Department of Science and Technology, New Delhi, and infrastructure support from the Centre for Nanotechnology (University of Hyderabad) are gratefully acknowledged.

REFERENCES

[1] C. L. Brown, G. Bushell, M. W. Whitehouse, D. S. Agrawal, S. G. Tupe, K. M. Paknikar, E. R. T. Tiekink, *Gold Bull.* **2007**, *40*, 245.
[2] M. Faraday, *Philos. Trans. R. Soc. Lond.* **1857**, *147*, 145.
[3] D. L. Feldheim, C. A. Foss Jr. (eds) *Metal Nanoparticles: Synthesis, Characterization and Applications.* Marcel Dekker: New York, Basel, 2002.
[4] P. K. Jain, X. Huang, I. H. El-Sayed, M. A. El-Sayed, *Acc. Chem. Res.* **2008**, *41*, 1578.
[5] R. Sardar, A. M. Funston, P. Mulvaney, R. W. Murray, *Langmuir* **2009**, *25*, 13840.
[6] G. V. Ramesh, S. Porel, T. P. Radhakrishnan, *Chem. Soc. Rev.* **2009**, *38*, 2646.
[7] F. Boneta, C. Guéry, D. Guyomard, R. H. Urbina, K. Tekaia-Elhsissen, J.-M. Tarascon, *Int. J. Inorg. Mater.* **1999**, *1*, 47.
[8] S. Porel, S. Singh, S. S. Harsha, D. N. Rao, T. P. Radhakrishnan, *Chem. Mater.* **2005**, *17*, 9.
[9] S. Clémenson, L. David, E. Espuche, *J. Polym. Sci. A* **2007**, *45*, 2657.
[10] T. W. H. Oates, E. Christalle, *J. Phys. Chem. C* **2007**, *111*, 182.
[11] S. Porel, N. Venkatram, D. N. Rao, T. P. Radhakrishnan, *J. Appl. Phys.* **2007**, *102*, 033107.

[12] E. Hariprasad, T. P. Radhakrishnan, *Chem. Eur. J.* **2010**, *16*, 14378.

[13] G. V. Ramesh, B. Sreedhar, T. P. Radhakrishnan, *Phys. Chem. Chem. Phys.* **2009**, *11*, 10059.

[14] S. Porel, S. Singh, T. P. Radhakrishnan, *Chem. Commun.* **2005**, 2387.

[15] S. Porel, N. Hebalkar, B. Sreedhar, T. P. Radhakrishnan, *Adv. Funct. Mater.* **2007**, *17*, 2550.

[16] B. Karthikeyan, M. Anija, P. Venkatesan, C. S. S. Sandeep, R. Philip, *Opt. Commun.* **2007**, *280*, 482.

[17] Y. A. Kumzerov, A. A. Nebereznov, S. B. Vakhrushev, B. N. Savenko, *Phys. Rev. B* **1995**, *52*, 4772.

[18] B. F. Borisov, E. V. Charnaya, P. G. Plotnikov, W. D. Hoffmann, D. Michel, Y. A. Kumzerov, C. Tien, C. S. Wur, *Phys. Rev. B* **1998**, *58*, 5329.

[19] L. Katsikas, M. Guttiérrez, A. Henglein, *J. Phys. Chem.* **1996**, *100*, 11203.

[20] A. Henglein, C. Brancewicz, *Chem. Mater.* **1997**, *9*, 2164.

[21] G. V. Ramesh, M. D. Prasad, T. P. Radhakrishnan, *Chem. Mater.* **2011**, *23*, 5231.

[22] J. A. Creighton, D. G. Eadon, *J. Chem. Soc., Faraday Trans.* **1991**, *87*, 3881.

[23] C. S. Barrett, *Acta Crystallogr.* **1957**, *10*, 58.

[24] Ph. Buffat, J. -P. Borel, *Phys. Rev. A* **1976**, *13*, 2287.

[25] S. L. Lai, J. Y. Guo, V. Petrova, G. Ramanath, L. H. Allen, *Phys. Rev. Lett.* **1996**, *77*, 99.

[26] E. A. Olson, M. Yu. Efremov, M. Zhang, Z. Zhang, L. H. Allen, *J. Appl. Phys.* **2005**, *97*, 034304.

[27] N. Dave, M. Y. Chan, P. J. Huang, B. D. Smith, J. Liu, *J. Am. Chem. Soc.* **2010**, *132*, 12668.

[28] H. Shi, S. Liu, H. Sun, W. Xu, Z. An, J. Chen, S. Sun, X. Lu, Q. Zhao, W. Huang, *Chem. Eur. J.* **2010**, *16*, 12158.

[29] B. Adhikari, A. Banerjee, *Chem. Mater.* **2010**, *22*, 4364.

[30] R. K. Bera, A. K. Das, C. R. Raj, *Chem. Mater.* **2010**, *22*, 4505.

[31] G. V. Ramesh, T. P. Radhakrishnan, *ACS Appl. Mater. Interfaces* **2011**, *3*, 988.

[32] C. A. Witham, W. Huang, C. Tsung, J. N. Kuhn, G. A. Somorjai, F. D. Toste, *Nat. Chem.* **2010**, *2*, 36.

[33] S. Shylesh, V. Schünemann, W. R. Thiel, *Angew. Chem. Int. Ed.* **2010**, *49*, 3428.

[34] B. Nowack, H. F. Krug, M. Height, *Environ. Sci. Technol.* **2011**, *45*, 1177.

[35] S. Porel, D. Ramakrishna, E. Hariprasad, A. D. Gupta, T. P. Radhakrishnan, *Curr. Sci.* **2011**, *101*, 927.

[36] B. Karthikeyan, *Physica B* **2005**, *364*, 328.

[37] B. Karthikeyan, M. Anija, R. Philip, *App. Phys. Lett.* **2006**, *88*, 053104.

[38] X. Meng, K. Fujita, Y. Zong, S. Murai, K. Tanaka, *Appl. Phys. Lett.* **2008**, *92*, 20112.

[39] D. Yu, W. Lin, C. Lin, L. Chang, M. Yang, *Mater. Chem. Phys.* **2007**, *101*, 93.

[40] G. V. Ramesh, K. Sudheendran, K. C. J. Raju, B. Sreedhar, T. P. Radhakrishnan, *J. Nanosci. Nanotechnol.* **2009**, *9*, 261.

[41] R. Abargues, J. Marqués-Hueso, J. Canet-Ferrer, E. Pedrueza, J. L. Valdés, E. Jiménez, J. P. Martínez-Pastor, *Nanotechnology* **2008**, *19*, 355308.

[42] J. Marqués-Hueso, R. Abargues, J. Canet-Ferrer, S. Agouram, J. L. Valdés, J. P. Martínez-Pastor, *Langmuir* **2010**, *26*, 2825.

[43] M. Bernabò, A. Pucci, F. Galembeck, C. A. de Paula Leite, G. Ruggeri, *Macromol. Mater. Eng.* **2009**, *294*, 256.

[44] S. Clémenson, E. Espuche, L. David, D. Léonard, *J. Membr. Sci. A* **2010**, *361*, 167.

7

PHOTOINDUCED GENERATION OF NOBLE METAL NANOPARTICLES INTO POLYMER MATRICES AND METHODS FOR THE CHARACTERIZATION OF THE DERIVED NANOCOMPOSITE FILMS

A. Pucci[1,2,3] and G. Ruggeri[1,2]

[1] *Macromolecular Science Group, Department of Chemistry and Industrial Chemistry, University of Pisa, Pisa, Italy*
[2] *INSTM, Unità di Ricerca di Pisa, Pisa, Italy*
[3] *CNR NANO, Istituto Nanoscienze-CNR, Pisa, Italy*

7.1 INTRODUCTION

Noble metal nanoparticles (MNPs) incorporated in polymeric matrices have been the object of considerable scientific efforts due to their potential applications as nanostructured materials [1–4].

Nanoparticles (NPs) embedded in an inert polymer matrix, which simply stabilizes the dispersion, provide enhanced and innovative catalytic activities as the external metal

Nanocomposites: In Situ *Synthesis of Polymer-Embedded Nanostructures*, First Edition.
Edited by Luigi Nicolais and Gianfranco Carotenuto.
© 2014 John Wiley & Sons, Inc. Published 2014 by John Wiley & Sons, Inc.

atoms of the NP can be considered as coordinatively unsaturated species. For example, bulk gold is chemically inert, whereas gold NPs have demonstrated to be extremely active in the oxidation of carbon monoxide [2, 5–7].

On the other side, the high efficiency of the aforementioned nanodispersion allows to provide the polymer matrix with unique optical properties without altering the thermoplastic behavior.

More specifically, depending on particle size, shape, and aggregation, NPs may confer tuneable absorption and scattering characteristics to the host matrices. At very low diameters (a few nm) and once dispersed into polymers in a non-aggregated form, NPs can lead to very low scattering materials, overcoming the very widely encountered problem of opacity of heterogeneous composites for optical applications [8–10].

The interesting optical attributes of MNPs are due to their unique interaction with light. Differently from smooth metal surfaces or metal powders, clusters of noble metals, such as gold, silver, or copper, assume a real and natural color due to the absorption of visible light at the surface plasmon resonance (SPR) frequency [7, 8, 10, 11]. Under the irradiation of light, the conduction electrons of the MNPs collectively oscillate at the resonance frequency providing the absorption of photons. Part of these photons are successively converted in phonons or in lattice vibrations, an effect associable to an absorption. As described by the Drude–Lorentz–Sommerfeld [8, 12] theory and shown by a huge number of experimental data [8, 13], the decrease in metal particle size leads to broadening of the absorption band, decrease of the maximum intensity, and often a hypsochromic (blue) shift of the peak and these effects depend also on cluster topology and packing [12, 14–16].

In addition, colloidal inorganic nanocrystals made of a few hundred up to a few thousand of atoms (II–VI groups) are receiving considerable attention due to their appealing properties (optical, electrical, and catalytic) derived from the zero-dimensional quantum-confined characteristics [17–21]. The color tuneability of both absorption and emission of those semiconducting nanocrystals as a function of size is actually one of the most attractive properties.

An even more expanding area of nanoscience is devoted to the study and preparation of innovative nanocomposite structures that allow the fabrication of nanocomposite-based devices, such as light-emitting diodes, photodiodes, and photovoltaic solar cells [22–24]. Noble MNPs or semiconducting nanocrystals efficiently incorporated into several polymeric matrices enhanced their optical properties (absorption, luminescence, and nonlinearity), thanks to the size and growth stabilization provided by the macromolecular support [23, 25–29].

The most common procedure to obtain a dispersion of MNPs in a polymer matrix is to prepare a colloidal solution of stabilized MNPs and then to mix it with the desired polymer in a mutual solvent and cast a film by evaporation from the solution [25, 30]. On the contrary, few examples are reported showing the dispersion of preformed MNPs in a polymer matrix by melt mixing at high temperature [31, 32].

Usually a water-soluble metal salt is moved in an organic solvent by using a tetraalkylammonium bromide as phase transfer agent and successively reduced with sodium borohydride in the presence of alkylthiols as surface stabilizers to prevent coalescence of growing NPs [33, 34].

In addition to thiols, different surface stabilizers have been used such as amines [35], poly(N-vinylpyrrolidone) (PVP) [36], and sodium poly(acrylic acid) [37]. By using the just described colloid-chemistry technique, MNPs have been dispersed in ultrahigh molecular weight poly(ethylene) [38, 39], high-density poly(ethylene) [32], poly(vinyl alcohol) [25, 40, 41], poly(methyl methacrylate) [42], poly(dimethylsiloxane) [43, 44], and poly(styrene-*b*-ethylene/propylene) [45].

Another approach for the preparation of nanocomposite containing MNPs involves the *in situ* formation of the NPs directly within the polymer matrix [40, 46–66]. This process is simple and just requires the reduction of the metal ion precursors by a photochemical or a thermal-induced process. However, differently from NPs prepared by the most common colloid-chemistry method, the control of the size distribution of the *in situ*-prepared particles resulted often to be more difficult to realize, due to the influence of several factors such as the polymer matrix composition and the time and the energy density of the photo- or thermo-irradiation process.

The present chapter reports methods developed for the preparation of noble metal nanocomposites by means of controlled photoinduced reduction processes starting from metal precursors. Reaction mechanisms, conditions, and nature of the polymer matrices are taken into account and discussed in terms of providing the control of the final NP morphology. Nanoparticle characteristics and composite properties are described by different analytical techniques.

7.2 PREPARATION OF NOBLE METAL–POLYMER NANOCOMPOSITES BY PHOTOREDUCTION PROCESS

7.2.1 Nature of the Polymer Matrix

Polymeric materials have been employed frequently as NP stabilizers during the chemical synthesis of metal colloids [67]. More specifically, several polymer matrices have been utilized as supporting and stabilizing materials for noble MNPs generated by photoinduced synthetic processes (Fig. 7.1).

For example, poly(vinyl alcohol) [57, 58, 69–71], poly(ethylene-co-vinyl alcohol) [70], poly(ethylene glycol) (PEG) [49, 50, 58, 64, 68], PVP [59, 66, 72], poly(diallyldimethylammonium chloride) [73], liposomes [74], and poly(oxyethylene)-*iso*-octylphenylether (Triton X-100) [55, 75] are reported to efficiently protect MNPs against aggregation and precipitation, thanks to the effective steric stabilization provided by the macromolecular support. Moreover, kinetic (electrostatic) or even more efficient thermodynamic stabilization [76] is also operated by the different functional groups composing the repeating units of the polymer matrices. Since most of them are water soluble, MNPs are generally formed or embedded within polymers starting from procedures that involve the use of water, polar (and also protic) solvents, or their mixtures.

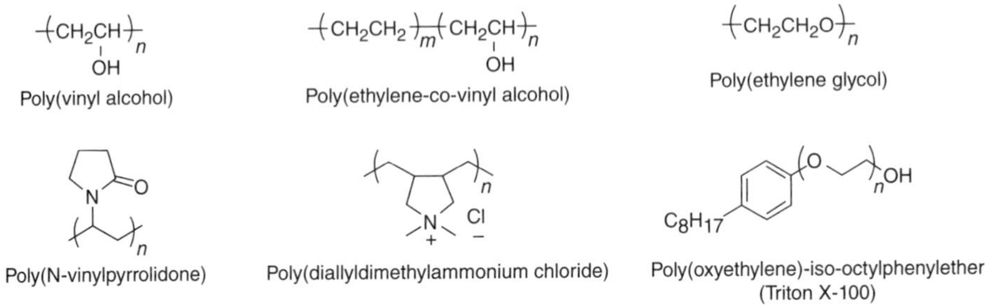

Figure 7.1. Polymeric materials used for the stabilization of MNPs.

7.2.2 Nature of the Metal

Silver. H. Hada et al. [77] reported for the first time the photoreduction of silver ion in aqueous and alcoholic solutions. According to the authors, the photolysis of silver perchlorate in water that occurred at about 254 nm proceeds through the electron transfer from the water molecule to silver ion according to Equations. 7.1 and 7.2:

$$\left(Ag^+, H_2O\right)_{cage} + h\nu \rightleftharpoons \left(Ag, H^+, OH\right)_{cage} \tag{7.1}$$

$$\left(Ag, H^+, OH\right)_{cage} \rightarrow Ag + H^+ + OH \tag{7.2}$$

Ag atoms are surrounded by the cage composed of water molecules and their decomposition products (OH radicals). The solvent cage gradually decomposes, thus converting the silver atoms in larger NPs.

Silver atoms are also generated in alcoholic solution following a similar mechanism (Equation 7.3):

$$\left(Ag^+, RCH_2OH\right)_{cage} \rightarrow Ag + H^+ + R\dot{C}HOH \tag{7.3}$$

Moreover, the alcoholic radical can react with another silver ion, thus converting to an aldehyde (Equation 7.4):

$$R\dot{C}HOH + Ag^+ \rightarrow Ag + H^+ + RCHO \tag{7.4}$$

The authors reported that the reaction rate is influenced by the nature of the solvent. In particular the rate increases in the order 2-methyl-2-propanol << water < ethanol < 2-propanol due to the ease of abstraction of α-hydrogen radical from these molecules. In a later paper [78], the authors confirmed the mechanism proposed for silver photoreduction in water and demonstrated the usefulness of protective agents such as liposomes for loading and stabilizing ultrafine noble metal particles against reoxidation and/or agglomeration of the latter.

The same research group [78] and A. Henglein [79] reported the photochemical formation of silver colloids by using acetone as solvent. Differently from mechanisms

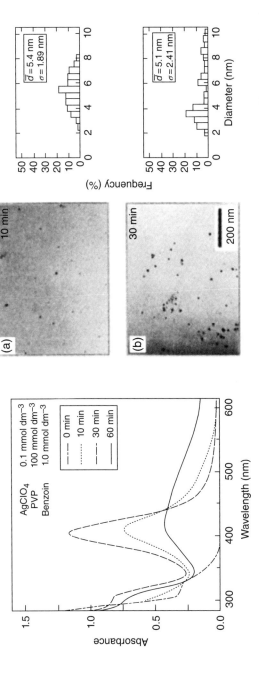

Figure 7.2. UV–Vis absorption spectra of $AgClO_4$ ethanol solutions containing benzoin with UV irradiation time (left). Transmission electron microscopy micrographs and size distributions of silver NPs after 10 and 30 min of irradiation (right). Adapted with permission from Reference 59. Copyright (1995) American Chemical Society.

described in water and in alcoholic solutions, acetone ketyl radical generated by ultraviolet (UV) excitation were reported to promote peptization of silver agglomerates generated in the first step of the photoinduced process, therefore providing fine silver particles stabilized by sodium dodecyl sulfate.

The first example of polymers as stabilizers of silver NPs was reported by K. Esumi et al. [59] who prepared ultrafine copper, gold, and silver NPs by UV irradiation of their metal salts dissolved in ethanol in the presence of PVP. The photoreduction was assisted by using benzoin as a photoinitiator, which generates benzoin radicals when irradiated by UV, thus promoting metal ion reduction.

The formation of silver NPs in ethanol solution was detected by UV–Vis absorption spectroscopy analyses, monitoring the evolution of the typical surface plasmon absorption band centered for silver clusters at about 400 nm (Fig. 7.2, left) [8].

The SPR band of nanostructured metals, described by the Drude–Lorentz–Sommerfeld theory [80], is strongly influenced by particle size, topology, and packing extent and enables the effective determination of any phenomenon influencing NP characteristics [7, 13, 14, 81].

In Figure 7.2, the SPR band of silver increased in the absorbance with UV irradiation time, flanked by a blue shift from 410 nm to 404 nm in the first 30 min of reaction. Conversely, the broad and redshifted absorption that occurred after 1 h of irradiation denoted strong particle aggregation. At the early stage of the reduction, the modification of the position of the absorption maximum was attributed by the authors to the change in the density of the conduction electron of the metal due to the injection of electrons into silver NPs by benzoin radicals. The fact that the position of the SPR band could be also attributed to particles aggregation was rejected by results coming from electron microscopies (Fig. 7.2, right). Transmission electron microscopy (TEM) experiments performed at different irradiation times (after 10 and 30 min of reduction, respectively) showed that the average diameters of silver NPs at the initial and final stages (namely, not considering samples at irradiation time higher than 30 min) of the reduction are nearly equal without any aggregation. Interestingly, with increasing benzoin concentration, a combination of effect occurs: benzoin radicals are able to nucleate smaller silver particles, and the increased quantity of injected electrons promoted the formation of highly dispersed silver particles due to strong electrical repulsion forces.

M. S. Scurrell et al. [49] proposed a simple photochemical method for the preparation of silver NPs in water in the presence of PEG. The polymer is reported to act both as source of free radicals and as a stabilizer, which is able to restrict the particle growth. More specifically, energy transfer throughout the solution ensures the homogeneous distribution of the photolytic radicals formed by the excitation and ionization of water. Then, PEG dissolved in solution behaves as a scavenger for the photolytic radicals, thus forming, due to the abstraction of α-hydrogens, a macroradical that is a very powerful reducing agent for silver (Scheme 1).

As a matter of fact, the macroradical facilitates the formation of silver NPs from ionic precursors, and the excess of PEG also serves as the stabilizing agent against silver NP aggregation and precipitation. The formation of Ag NPs was monitored by UV–Vis spectroscopy and confirmed by TEM. More specifically, the electron diffraction pattern [82] caused by the scattering of electron that passed through the specimen was utilized

$$H_2O \xrightarrow{h\nu} H^{\bullet} + OH^{\bullet}$$

$$(CH_2CH_2O)_n\, CH_2CH_2OH \xrightarrow{H^{\bullet}} (CH_2CH_2O)_n\, CH_2\overset{\bullet}{C}HOH + H_2$$

$$(CH_2CH_2O)_n\, CH_2CH_2OH \xrightarrow{OH^{\bullet}} (CH_2CH_2O)_n\, CH_2\overset{\bullet}{C}HOH + H_2O$$

$$(CH_2CH_2O)_n\, CH_2\overset{\bullet}{C}HOH \xrightarrow{Ag^+} Ag^0 + (CH_2CH_2O)_n\, CH_2CHO + H^+$$

Scheme 1. Photoinduced reduction processes that occur during the formation of PEG-stabilized silver NPs.

to determine the crystalline (metallic) nature of the examined particles (Fig. 7.3, left). On the other hand, the energy-dispersive spectrum [82] (EDS, Fig. 7.3, right), recorded by collecting and counting the X-rays emitted from the specimen surface, permitted the identification of silver in the NPs.

The size of silver NPs was modulated as a function of irradiation time and PEG concentration: larger silver NPs were obtained for longer reaction time and more diluted PEG solutions.

The influence of the morphology of silver colloids on their optical properties was reported by A. Callegari et al. [53] that prepared photochemically grown silver NPs via a wavelength-controlled process. More specifically, spherical silver NPs were firstly produced by sodium borohydrate reduction of $AgNO_3$ dissolved in water in the presence of PVP as protecting agent. The colloidal dispersion was then divided into several identical samples that were exposed for about 1 h to the same light source (350 nm < λ < 700 nm). After this induction period, each sample is assigned a different illumination condition by fitting in front of it a specific colored glass filter, and the reaction was then allowed to proceed for several hours.

Interestingly, the authors compared the UV–Vis absorption spectra of the different excited samples with the TEM micrographs (Fig. 7.4), evidencing a strict correlation between the position of the SPR band and NP morphology.

All the initially prepared colloidal solution exhibited a prominent peak in the 400 nm region, corresponding to the plasmon resonance of the small spherical particles (black curve). After the induction period of exposition to the fluorescent tube (blue curve), the 400 nm plasmon peak slightly increased in size and width generating a shoulder at 450 nm (attributed to small particles and aggregates of them with small aspect ratio) and was flanked by the appearance of a small peak at about 630 nm (attributed to plates with large aspect ratio) [10, 83].

Conversely, in the subsequent period, these species evolved differently, depending on the specific illumination conditions. The use of blue filters allowed the growth of the peak at about 430 nm, whereas the progressively use of longer-pass orange filters promoted the decreasing of the band at 430 nm and the evolution of even more redshifted absorptions higher than 630 nm. Transmission electron microscopy micrographs confirmed the growth of the different silver nanostructures as small NP aggregates or plates with large aspect ratio depending on the wavelength of the excitation. Moreover, the

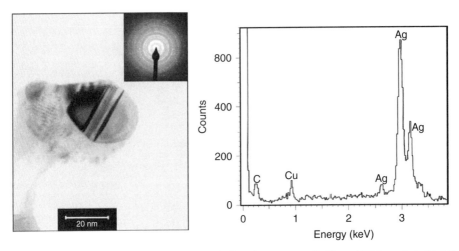

<u>Figure 7.3.</u> High-magnification TEM with diffraction pattern (left) and EDS analysis (right) showing the formation of silver NPs. Carbon and copper peak is from the carbon-coated copper sample grid. Reprinted with kind permission from Springer Science+Business Media: Reference 49.

preparation of nanostructured silver particles with absorption in the near infrared (about 800 nm) could allow the application of these structures as biomarkers [84].

Recently, L. Balan and J.-P. Malval reported the preparation of silver NPs embedded in a PEG-based polymer through a photoinduced one-step method [64]. In detail, silver precursor (1% by wt.) was dispersed in a poly(ethylene glycol) diacrylate monomer with a synthesized 2,7-diaminofluorene (DAF) derivative (0.5% by wt.) that acted both as photoreductant for the metal and as photoinitiator for the monomer. Moreover, when N-methyl diethanolamine (MDEA) as co-initiator (1% by wt.) was added, a clear acceleration of the photopolymerization rate occurred, and, according to the stabilizing effect towards MNPs provided by amines, nanocomposite polymer exhibited a uniform distribution of fine silver particles with 4–5 nm in average diameter as revealed by TEM micrograph (Fig. 7.5).

In detail, after the excitation of the DAF derivative at 365 nm with Xe lamp, a radical cation DAF$^{\cdot+}$ is generated that then reacts with MDEA according to the following mechanism:

$$\begin{aligned} DAF^{\cdot+} + MDEA &\rightarrow DAFH^+ + MDEA^{\cdot} \\ MDEA^{\cdot} + Ag^+ &\rightarrow MDEA^+ + Ag^0 \end{aligned} \tag{7.5}$$

Finally, the amine radicals donate an electron to silver ions that are converted into Ag NPs that are efficiently coated by the MDEA coupling effect.

Another interesting example of Ag NPs generated directly within a polymer matrix was presented by Bernabò et al. [71] The authors reported that nanocomposites based on PVA and silver NPs were efficiently generated by a photoinduced reduction process promoted by the exposition of the precursor directly to sunlight. In this case, ethylene

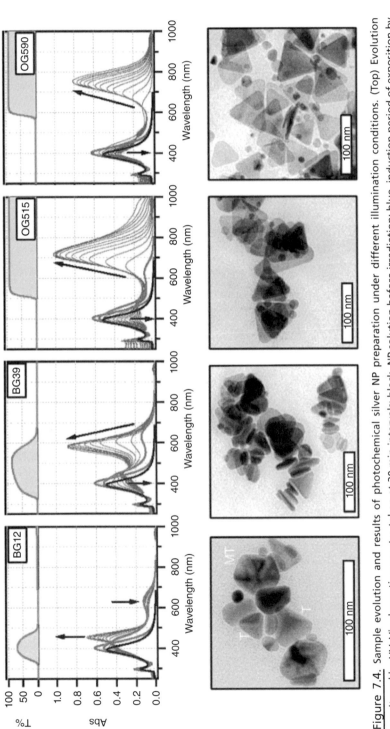

Figure 7.4. Sample evolution and results of photochemical silver NP preparation under different illumination conditions. (Top) Evolution monitored by UV–Vis absorption spectra taken at 30 min intervals: black, NP solution before irradiation; blue, induction period of exposition by using a conventional fluorescent tube; red, second stage of growth by sample illumination fitted with a specific colored filter. The transmission of the filters used is indicated by the green shaded area on top of each graph. The arrows indicate the evolution of the main spectral features during this stage. (Bottom) TEM micrographs of the resulting silver NPs. Reprinted with permission from Reference [53]. Copyright (2003) American Chemical Society. (See insert for color representation of the figure.)

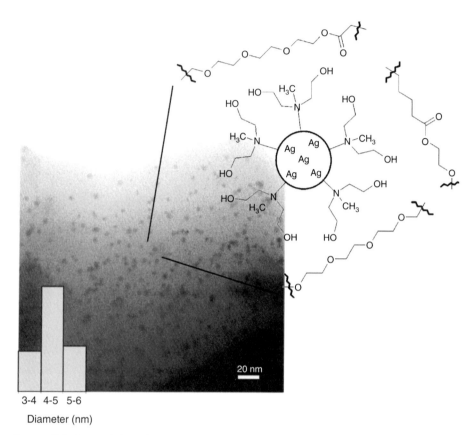

3-4 4-5 5-6

Diameter (nm)

Figure 7.5. Transmission electron microscopy images of Ag NPs embedded in PEG matrix, with their respective size distribution and schematic model of the amine-capped Ag NPs. Adapted with permission from Reference [64]. Copyright (2008) American Chemical Society.

glycol was mixed within the polymer matrix containing Ag⁺ precursor and used as electron source because it is rapidly oxidized by UV sunlight giving the electron necessary for the Ag⁰ NP formation [57]. Accordingly, the samples appeared totally colorless before exposure, but they became yellowish just after 5 min of irradiation, turning gradually more intense with increasing irradiation time showing the typical surface plasmon absorption band at about 435 nm (Fig. 7.6).

The bright-field TEM micrograph of the PVA nanocomposite film confirmed the formation of NPs within the polymer matrix with approximately spherical and oblate spheroid shape and a mean size of about 13 nm (Fig. 7.7a). Moreover, the selected area diffraction (SAD) pattern of the nanostructures confirmed the chemical composition of NPs that resulted based on Ag⁰ (Fig. 7.7b).

Interestingly, the formation of prolate NPs (aspect ratio of about 1.7) during sun irradiation was attributed by the authors to the presence in the PVA matrix of the excess of ethylene glycol that can act as a sort of particle shape director, thus promoting the

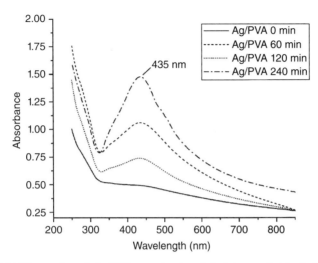

Figure 7.6. UV–Vis spectra of Ag/PVA samples irradiated by sunlight at different times. Adapted from Reference [71], with permission from John Wiley and Sons.

anisotropic growth of MNPs. The crystalline (metallic) nature of the Ag NPs generated by the sunlight photoreduction process was demonstrated by X-ray diffraction (XRD) analysis (Fig. 7.7c), which revealed the characteristic reflections of Ag NPs with face-centered cubic (fcc) geometry. Conversely, the diffuse broad contribution at small angle was instead attributed to the semicrystalline PVA matrix, corresponding to the (101) plane of polymer crystals [70].

Gold. One of the first works concerning the photogeneration of Au NPs within polymers was proposed by G. Mills et al. [73] which described the photoinitiated formation of Au NPs in polymer gels. Au NPs were formed by photoreducing $AuCl_4^-$ ions that were incorporated into methanol-swollen cross-linked poly(diallyldimethylammonium chloride) (see Fig. 7.1). Illumination of the gels by a UV source induced a reduction of the absorption band of gold precursor at 318 nm since $AuCl_4^-$ ions are photoreduced in alcohols to $AuCl_2^-$ and then to Au^0, presumably by disproportionation:

$$3AuCl_2^- \rightleftharpoons 2Au^0 + AuCl_4^- + 2Cl^- \tag{7.6}$$

The formation of Au NPs was confirmed by the emersion of the typical SPR band of gold at about 550 nm in UV–Vis spectrum and by XRD analysis.

A more complete contribution in the preparation of Au NPs by a photoinduced reduction process was reported by Z. Y. Chen [58] and I. Tanahashi [57] who described the photoinduced Au NPs formation by using poly(vinyl alcohol) as protecting layer for NPs. In the first case [58], Z. Y. Chen et al. reported the shape-controlled synthesis of gold NPs using a low-pressure mercury lamp (253.7 nm) at room temperature. The very slow UV irradiation photoreduction process favored the formation of shaped gold NPs. More specifically, they demonstrated that not only the concentration of Au precursor and the irradiation time but also the concentration of PVA play important roles in the

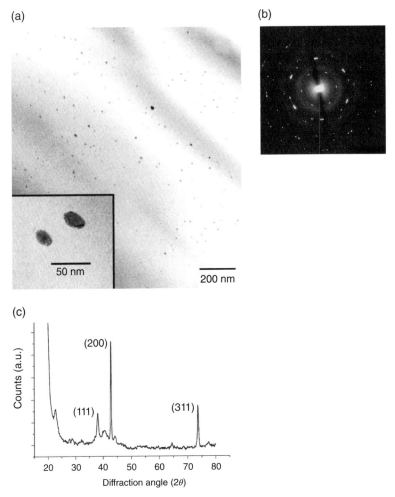

Figure 7.7. (a) Bright-field TEM micrograph of PVA nanocomposite and (b) SAD pattern of the structures evidenced in (a). (c) X-ray diffractogram of the nanocomposite. Adapted from Reference [71], with permission from John Wiley and Sons.

morphology control of the gold NPs. When PVA/Au³⁺ ratio was increased, the formation of the Au NPs with polyhedral shapes occurred.

In the second case, I. Tanahashi and H. Kanno [57] reported the temperature dependence of photoinduced Au NPs formation within PVA films. As reported by Bernabò et al. [71] in the case of silver and by M. A. El-Sayed [72], Au NPs were generated in the polymer matrix, thanks to the presence of ethylene glycol, which can be easily oxidized to aldehyde due to the reactivity of α-hydrogens. However, the PVA film containing the Au precursor did not show color variations up to 20 h of irradiation at 0 °C. Conversely, the formation of the photoinduced Au NPs became faster with increasing the reaction temperature (Fig. 7.8).

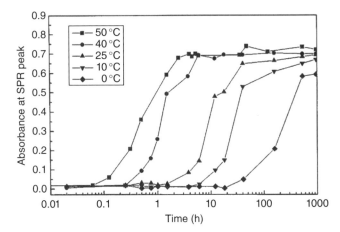

Figure 7.8. Intensity variation of the SPR peak of the Au/PVA films as a function of reaction time at different temperatures. Reprinted with permission from Reference [57]. Copyright 2000, American Institute of Physics.

In detail, the color of the film changed from yellow before irradiation to purple due to the SPR band of NPs.

The average diameter of Au NPs generated within PVA film was determined by TEM measurements to be about 30 nm with a large size distribution, likely due to the reduction process of Au^{3+} ions that occurred in the solid state. The authors estimated the mean diameter of Au NPs by XRD as well, by applying Scherrer's equation

$$d = \frac{0.9\lambda}{\beta \cos\theta} \tag{7.7}$$

where λ is the wavelength of the X-ray source and β the full width at half maximum of the XRD peak at the diffraction angle θ. In the XRD pattern (Fig. 7.9), the film before irradiation (a) showed the typical amorphous halo pattern around $2\theta = 20°$ due to the supporting glass substrate. Conversely, peaks appeared after the irradiation (b) at $2\theta = 38.1°$ and $44.4°$, respectively, attributed to the (111) and (200) reflection lines of cubic Au.

The mean diameter of Au NPs was determined from the line broadening of the diffraction band at $2\theta = 38.1°$ to be about 30 nm, in line with the data calculated by TEM investigations.

The synthesis of metal NPs of desired size and shape resulted therefore to be extremely important because of their size and morphological-dependent optoelectronic properties. T. Pal et al. [55] proposed the use of a seed-mediated successive growth process by UV irradiation for the preparation of monodispersed Au NPs of controlled size. The "seed particles" of about 18 nm were prepared by UV irradiation in water in the presence of Triton X-100 as particle stabilizer (Fig. 7.1). The irradiation promoted the photolysis of water into radicals that were then captured by Triton X-100 that acts also as a scavenger, thus promoting the reduction of Au^{3+} precursor. The seed particles were then grown to the larger size up to 70 nm by the successive addition of metal ions and

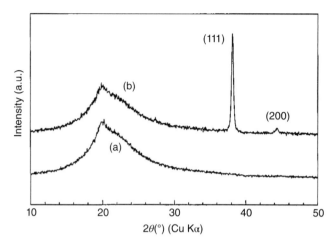

Figure 7.9. X-ray diffraction patterns of the Au/PVA film on a glass substrate before (a) and after (b) UV irradiation at room temperature. Reprinted with permission from Reference [57]. Copyright 2000, American Institute of Physics.

again UV irradiation under the same experimental condition. Depending on the concentration of the Au^{3+} precursor added in the second irradiation step, the average size of the final NP can be easily modulated.

The "seed particles" showed an SPR band at about 530 nm (Fig. 7.10) that gradually broadened towards higher wavelengths according to the growth of the gold particles [8] in solution that occurred after the second irradiation step.

A similar photochemical synthesis of colloidal gold particles was proposed by H. Zhao et al. [50] and based on a seed-mediated growth approach under the UV solar radiation. More specifically, the $HAuCl_4$ (Au^{3+}) solution containing certain amounts of protective agent as PEG and acetone was exposed for 10 min to a UV 300 nm irradiation, and Au "seed" particles with 5 nm in average diameter were obtained. Their further growth then was carried out in the identical constituent solutions under the solar radiation profiting of its UV region from 280 nm to 320 nm, thus allowing the preparation of size-controlled Au NPs in larger scale.

According to the authors, the reduction of [$AuCl_4^-$–PEG] associated complex was greatly accelerated in the presence of acetone:

$$(CH_3)_2 CO \xrightarrow{h\nu} CH_3COCH_3^*$$

$$CH_3COCH_3^* + H^+ \rightarrow (CH_3)_2 \overset{\cdot}{C}OH \qquad (7.8)$$

$$(CH_3)_2 \overset{\cdot}{C}OH + [AuCl_4^- -PEG] \rightarrow [Au^0 -PEG] + (CH_3)_2 CO + 4Cl^-$$

Being similar to photochemical formation of Ag NPs [78, 79], acetone ketyl radical produced by excitation of acetone by UV light plays an accelerating role in the reduction of Au^{3+} ions, thus resulting in a faster nucleation rate.

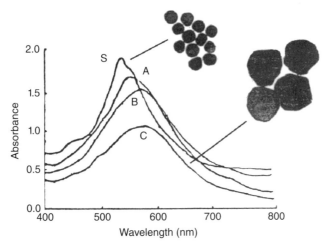

Figure 7.10. UV–Vis spectra of Au NP solutions with different sizes. S represents the seed NP sol; A, B, and C represent samples of increasing particle size as revealed by TEM micrographs. Adapted from Reference [55], with permission from Elsevier.

An interesting method aimed to the controlled preparation of Au NPs with various average size was proposed by A. Pucci et al., which described the photoinduced formation of gold NPs within vinyl alcohol-based polymer films [70]. More specifically, the UV illumination initially generated gold NPs in the presence of ethylene glycol [72] with an average diameter ranging from 12 nm to 23 nm, depending on the nature of the polymeric film. On passing from poly(vinyl alcohol) to poly(ethylene)-co-(vinyl alcohol) (EVAl, Fig. 7.1) composed by different amounts of apolar and in principle more stable ethylene units (27% and 44% by mol, respectively), larger Au NPs were produced from the dissolved Au^{3+} precursor due to the lower –OH content of the copolymers and hence the reduced possibility of stabilizing the colloidal interface.

On the other hand, a longer irradiation time caused the generation of a large number of smaller gold NPs, lowering their average particle size down to about 5 nm. This phenomenon was attributed by the authors to the fact that due to the decreased concentration of Au at long irradiation time (>30 min), the nucleation rate became more and more competitive with the growth rate.

In Fig. 7.11, some of the characterization methods reported by the authors were pictured. Transmission electron microscopy micrograph evidences the presence of a relevant number of spherical Au NPs with average size well below 10 nm (Fig. 7.11a). Moreover, the XRD (Fig. 7.11b) showed the appearance of diffraction peaks after UV illumination, which can be assigned to the fcc unit cell of gold. Interestingly, the authors described that with increasing irradiation time, the signal to noise ratio of the Au peaks improves indicating the progressive formation of nanostructured metallic Au^0 during the photoreduction process. Atomic force microscopy (AFM) analysis, a useful tool for the submicro-characterization of material surfaces, clearly confirmed the presence of uniform very small Au NPs also close to the air-exposed polymer surface (Fig. 7.11c).

(a)

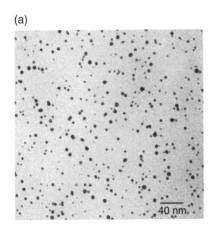

(b)

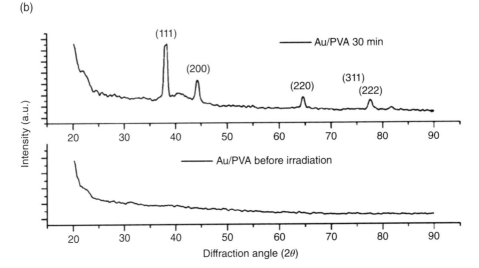

(c)

Figure 7.11. (a) Bright-field TEM micrograph of Au NPs/PVA film irradiated for 30 min; (b) XRD patterns of Au NPs/PVA films before and after UV irradiation for 30 min; and (c) 8 × 8 μm² AFM deflection image of the same irradiated sample film. Adapted from Reference [70], with permission from Royal Society of Chemistry.

7.3 CONCLUSIONS

The results presented and discussed in the present chapter provide a clear demonstration about the potentiality and advantages offered by using the photochemical method for the preparation of size-controlled noble MNPs by using polymers as effective stabilizing layer. More specifically, the controlled reduction of metal ions (silver and gold are the most investigated) can be carried out without using excess of reducing agent, and the rate of reaction is well controlled, since the number of reducing equivalents generated by radiation is well defined and radiation is absorbed regardless of the presence of light-absorbing solutes and products. The reaction mechanisms, the process conditions, and the nature of the polymer matrices have proved as decisive factor for their ability to control and modulate the noble MNPs' average size and morphology. For this purpose, in this review, methods aimed for the determination of the photogenerated NPs' characteristics were thoroughly presented and illustrated.

Considering the general validity of the approach and the easiness of the process, methods concerning the direct formation of MNPs within a polymer matrix in the form of thin film appear as the most promising for the preparation of nanocomposites for several applications.

ACKNOWLEDGMENTS

Partial financial support by MIUR-PRIN 2008 2008SXASBC is kindly acknowledged.

REFERENCES

[1] Ciardelli, F.; Coiai, S.; Passaglia, E.; Pucci, A.; Ruggeri, G. *Polymer International* **2008**, *57*, 805–836.

[2] Klabunde, K. J. *Nanoscale Materials in Chemistry*; John Wiley & Sons, Inc.: New York, 2001.

[3] Heilmann, A. *Polymer Films with Embedded Metal Nanoparticles*; Springer-Verlag: Berlin, 2003.

[4] Nicolais, L.; Carotenuto, G. *Metal-Polymer Nanocomposites*; John Wiley & Sons: Hoboken, New Jersey, 2005.

[5] Sardar, R.; Funston, A. M.; Mulvaney, P.; Murray, R. W. *Langmuir* **2009**, *25*, 13840–13851.

[6] Schmid, G.; Corain, B. *European Journal of Inorganic Chemistry* **2003**, 3081–3098.

[7] Jain, P. K.; Huang, X.; El-Sayed, I. H.; El-Sayed, M. A. *Accounts of Chemical Research* **2008**, *41*, 1578–1586.

[8] Kreibig, U.; Vollmer, M. *Optical Properties of Metal Clusters*; Springer-Verlag: Berlin, 1995.

[9] Kamat, P. V. *Journal of Physical Chemistry B* **2002**, *106*, 7729–7744.

[10] Schwartzberg, A. M.; Zhang, J. Z. *Journal of Physical Chemistry C* **2008**, *112*, 10323–10337.

[11] Gehr, R. J.; Boyd, R. W. *Chemistry of Materials* **1996**, *8*, 1807–1819.

[12] Klabunde, K. J.; Editor *Nanoscale Materials in Chemistry*, 2001.

[13] Daniel, M.-C.; Astruc, D. *Chemical Reviews* **2004**, *104*, 293–346.

[14] Kelly, K. L.; Coronado, E.; Zhao, L. L.; Schatz George, C. *Journal of Physical Chemistry B* **2003**, *107*, 668–677.

[15] Hu, M.; Chen, J.; Li, Z.-Y.; Au, L.; Hartland, G. V.; Li, X.; Marquez, M.; Xia, Y. *Chemical Society Reviews* **2006**, *35*, 1084–1094.

[16] Wilcoxon, J. P.; Abrams, B. L. *Chemical Society Reviews* **2006**, *35*, 1162–1194.

[17] Murphy, C. J. *Analytical Chemistry* **2002**, *74*, 520A-526A.

[18] Sakurai, H. *Organometallic News* **2004**, 100.

[19] Willner, I.; Willner, B. *Pure and Applied Chemistry* **2002**, *74*, 1773–1783.

[20] Murakata, T.; Higashi, Y.; Yasui, N.; Higuchi, T.; Sato, S. *Journal of Chemical Engineering of Japan* **2002**, *35*, 1270–1276.

[21] Schmid, G. *Nanoparticles: From Theory to Application*; Wiley-VCH Verlag GmbH & Co.: Weinheim, 2004.

[22] Cozzoli, P. D.; Pellegrino, T.; Manna, L. *Chemical Society Reviews* **2006**, *35*, 1195–1208.

[23] Godovsky, D. Y. *Advances in Polymer Science* **2000**, *153*, 163–205.

[24] Trindade, T.; O'Brien, P.; Pickett, N. L. *Chemistry of Materials* **2001**, *13*, 3843–3858.

[25] Caseri, W. *Macromolecular Rapid Communications* **2000**, *21*, 705–722.

[26] Porel, S.; Singh, S.; Harsha, S. S.; Rao, D. N.; Radhakrishnan, T. P. *Chemistry of Materials*. **2005**, *17*, 9–12.

[27] Pucci, A.; Ruggeri, G.; Bronco, S.; Bertoldo, M.; Cappelli, C.; Ciardelli, F. *Progress in Organic Coatings* **2007**, *58*, 105–116.

[28] Pucci, A.; Ruggeri, G.; Ciardelli, F. In *Advanced Nanomaterials*; Geckeler, K., Nishide, H., Eds.; Wiley-VCH: Weinheim, 2010; Vol. 1, p 379–401.

[29] Pucci, A.; Boccia, M.; Galembeck, F.; Leite, C. A. d. P.; Tirelli, N.; Ruggeri, G. *Reactive & Functional Polymers*. **2008**, *68*, 1144–1151.

[30] Caseri, W. *Chemistry of Nanostructured Materials* **2003**, 359–386.

[31] Dirix, Y.; Bastiaansen, C.; Caseri, W.; Smith, P. *Journal of Materials Science* **1999**, *34*, 3859–3866.

[32] Dirix, Y.; Bastiaansen, C.; Caseri, W.; Smith, P. *Advanced Materials* **1999**, *11*, 223–227.

[33] Brust, M.; Walker, M.; Bethell, D.; Schiffrin, D. J.; Whyman, R. *Journal of the Chemical Society, Chemical Communications* **1994**, 801–802.

[34] Lu, H.; Lu, G. H.; Kessinger, A. M.; Foss, C. A. *Journal of Physical Chemistry B* **1997**, *101*, 9139–9142.

[35] Leff, D. V.; Brandt, L.; Health, J. R. *Langmuir* **1996**, *12*, 4723–4730.

[36] Carotenuto, G. *Applied Organometallic Chemistry* **2001**, *15*, 344–351.

[37] Hussain, I.; Brust, M.; Papworth, A. J.; Cooper, A. I. *Langmuir* **2003**, *19*, 4831–4835.

[38] Pucci, A.; Elvati, P.; Ruggeri, G.; Liuzzo, V.; Tirelli, N.; Isola, M.; Ciardelli, F. *Macromolecular Symposia* **2003**, *204*, 59–70.

[39] Pucci, A.; Tirelli, N.; Willneff, E. A.; Schroeder, S. L. M.; Galembeck, F.; Ruggeri, G. *Journal of Materials Chemistry* **2004**, *14*, 3495–3502.

[40] Perez-Juste, J.; Rodriguez-Gonzalez, B.; Mulvaney, P.; Liz-Marzan, L. M. *Advanced Functional Materials* **2005**, *15*, 1065–1071.

[41] Mbhele, Z. H. *Chemistry of Materials* **2003**, *15*, 5019–5024.

[42] Nakao, Y. *Journal of Colloid and Interface Science* **1995**, *171*, 386–391.

[43] Uhlenhaut, D. I.; Smith, P.; Caseri, W. *Advanced Materials* **2006**, *18*, 1653–1656.

[44] Pastoriza-Santos, I.; Perez-Juste, J.; Kickelbick, G.; Liz-Marzan, L. M. *Journal of Nanoscience and Nanotechnology* **2006**, *6*, 453–458.

[45] Bockstaller, M. R. *Journal of Physical Chemistry B* **2003**, *107*, 10017–10024.

[46] Mallick, K.; Witcomb, M. J.; Scurrell, M. S. *European Physical Journal: Applied Physics* **2005**, *29*, 45–49.

[47] Salvati, R.; Longo, A.; Carotenuto, G.; De Nicola, S.; Pepe, G. P.; Nicolais, L.; Barone, A. *Applied Surface Science* **2005**, *248*, 28–31.

[48] Stepanov, A. L. *Technical Physics (Translation of Zhurnal Tekhnicheskoi Fiziki)* **2004**, *49*, 143–153.

[49] Mallick, K.; Witcomb, M. J.; Scurrell, M. S. *Journal of Materials Science* **2004**, *39*, 4459–4463.

[50] Dong, S.; Tang, C.; Zhou, H.; Zhao, H. *Gold Bulletin* **2004**, *37*, 187–195.

[51] Kaneko, K.; Sun, H.-B.; Duan, X.-M.; Kawata, S. *Applied Physics Letters* **2003**, *83*, 1426–1428.

[52] Henneke, D. E.; Malyavanatham, G.; Kovar, D.; O'Brien, D. T.; Becker, M. F.; Nichols, W. T.; Keto, J. W. *Journal of Chemical Physics* **2003**, *119*, 6802–6809.

[53] Callegari, A.; Tonti, D.; Chergui, M. *Nano Letters* **2003**, *3*, 1565–1568.

[54] Seibel, M. In *PCT Int. Appl.*; (Plasco Ehrich Plasma-Coating G.m.b.H., Germany). Wo, 2001, p 12 pp.

[55] Mallick, K.; Wang, Z. L.; Pal, T. *Journal of Photochemistry and Photobiology, A: Chemistry* **2001**, *140*, 75–80.

[56] Gaddy, G. A.; McLain, J. L.; Steigerwalt, E. S.; Broughton, R.; Slaten, B. L.; Mills, G. *Journal of Cluster Science* **2001**, *12*, 457–471.

[57] Tanahashi, I.; Kanno, H. *Applied Physics Letters* **2000**, *77*, 3358–3360.

[58] Zhou, Y.; Wang, C. Y.; Zhu, Y. R.; Chen, Z. Y. *Chemistry of Materials* **1999**, *11*, 2310–2312.

[59] Itakura, T.; Torigoe, K.; Esumi, K. *Langmuir* **1995**, *11*, 4129–34.

[60] Porel, S.; Singh, S.; Radhakrishnan, T. P. *Chemical Communications* **2005**, 2387–2389.

[61] Ramesh, G. V.; Radhakrishnan, T. P. *ACS Applied Materials & Interfaces* **2011**, *3*, 988–994.

[62] Lollmahomed, F. B.; Narain, R. *Langmuir* **2011**, *27*, 12642–12649.

[63] Lee, S. J.; Piorek, B. D.; Meinhart, C. D.; Moskovits, M. *Nano Letters* **2010**, *10*, 1329–1334.

[64] Balan, L.; Jin, M.; Malval, J.-P.; Chaumeil, H.; Defoin, A.; Vidal, L. *Macromolecules* **2008**, *41*, 9359–9365.

[65] Tan, S.; Erol, M.; Attygalle, A.; Du, H.; Sukhishvili, S. *Langmuir* **2007**, *23*, 9836–9843.

[66] Han, M. Y.; Quek, C. H. *Langmuir* **2000**, *16*, 362–367.

[67] Pomogailo, A. D.; Kestelman, V. N. *Metallopolymer nanocomposites*; Springer-Verlag: Heidelberg, 2005.

[68] Longenberger, L.; Mills, G. *Journal of Physical Chemistry* **1995**, *99*, 475–478.

[69] Kaneko, K.; Sun, H. B. *Applied Physics Letters* **2003**, *83*, 1426–1428.

[70] Pucci, A.; Bernabò, M.; Elvati, P.; Meza, L. I.; Galembeck, F.; de Paula Leite, C. A.; Tirelli, N.; Ruggeri, G. *Journal of Materials Chemistry* **2006**, *16*, 1058–1066.

[71] Bernabò, M.; Pucci, A.; Galembeck, F.; de Paula Leite, C. A.; Ruggeri, G. *Macromolecular Materials and Engineering* **2009**, *294*, 256–264.

[72] Eustis, S.; Hsu, H. Y.; El-Sayed, M. A. *Journal of Physical Chemistry B* **2005**, *109*, 4811–4815.

[73] Weaver, S.; Taylor, D.; Gale, W.; Mills, G. *Langmuir* **1996**, *12*, 4618–4620.

[74] Sato, T.; Ito, T.; Iwabuchi, H.; Yonezawa, Y. *Journal of Materials Chemistry* **1997**, *7*, 1837–1840.

[75] Mallik, K.; Mandal, M.; Pradhan, N.; Pal, T. *Nano Letters* **2001**, *1*, 319–322.

[76] Cao, G. *Nanostructures and Nanomaterials: Synthesis, Properties, and Applications*; Imperial College Press: London, 2004.

[77] Hada, H.; Yonezawa, Y.; Yoshida, A.; Kurakake, A. *Journal of Physical Chemistry* **1976**, *80*, 2728–2731.

[78] Yonezawa, Y.; Sato, T.; Kuroda, S. *Journal of the Chemical Society, Faraday Transactions.* **1991**, *87*, 1905–1910.

[79] Henglein, A. *Chemistry of Materials* **1998**, *10*, 444–450.

[80] Quinten, M. *Optical Properties of Nanoparticle Systems: Mie and Beyond*; Wiley-VCH Verlag & Co. KGaA: Weinheim (Germany), 2011.

[81] Khlebtsov, N. G.; Trachuk, L. A.; Mel'nikov, A. G. *Optics and Spectroscopy* **2005**, *98*, 77–83.

[82] Goodhew, P. J.; Humphreys, J.; Beanland, R. *Electron Microscopy and Analysis*; Taylor & Francis: London, 2001.

[83] Halas, N. J.; Lal, S.; Chang, W. S.; Link, S.; Nordlander, P. *Chemical Reviews* **2011**, *111*, 3913–3961.

[84] Tang, L. C.; Dong, C. Q.; Ren, J. C. *Talanta* **2010**, *81*, 1560–1567.

8

INTERMATRIX SYNTHESIS AND CHARACTERIZATION OF POLYMER-STABILIZED FUNCTIONAL METAL AND METAL OXIDE NANOPARTICLES

A. Alonso[1], G.-L. Davies[3], A. Satti[3], J. Macanás[2], Y.K. Gun'ko[3], M. Muñoz[1], and D.N. Muraviev[1]

[1]*Universitat Autònoma de Barcelona, Barcelona, Spain*
[2]*Universitat Politècnica de Catalunya, Barcelona, Spain*
[3]*Trinity College Dublin, Dublin, Ireland*

8.1 INTRODUCTION

8.1.1 General Principles of Intermatrix Synthesis of Metal and Metal Oxide Nanoparticles in Functional Polymers

The development of nanoparticles (NPs) has been intensively pursued [1, 2]. Within this broad field, zerovalent metal NPs (MNPs) and metal oxide NPs (MONPs) have attracted particular interest for their known special properties in several scientific and technological areas [3–7]. In particular, MONPs are excellent candidates for understanding and controlling the magnetic properties of NPs through the variation of chemistry at the atomic level [8, 9]. The most reported magnetic compounds are magnetite (Fe_3O_4), cobalt ferrite ($CoFe_2O_4$), and manganese ferrite ($MnFe_2O_4$), which belong to the ferromagnetic oxide (or ferrite) family [10, 11]. The preparation of MNPs and MONPs can be carried out through various synthetic routes based on either bottom-up or top-down

Nanocomposites: In Situ *Synthesis of Polymer-Embedded Nanostructures*, First Edition.
Edited by Luigi Nicolais and Gianfranco Carotenuto.
© 2014 John Wiley & Sons, Inc. Published 2014 by John Wiley & Sons, Inc.

approaches, which have been summarized in recent publications [4, 5, 7, 9, 12–14]. One of the most frequently used procedures involves the use of capping stabilizing agents or surfactants, which help to prevent NP aggregation and Ostwald ripening [15, 16]. In such cases, stabilizers not only preserve NP size, but they also play a crucial role in controlling their shape and size [17].

Alternatively, the synthesis of nanocomposites (NCs) consisting of supporting polymer and inorganic NPs including MNPs or MONPs can be carried out by using various synthetic procedures. The most general classification of all these methods can be based on the conditions of the synthesis of the NPs, which can be carried out by using either *in situ* or *ex situ* synthetic procedures [3, 18]. By *in situ* methods, NPs can be grown inside a matrix using different techniques, yielding a material that can be directly used for a foreseen purpose [19–21]. In this case, the preparation of many metal–polymer and metal oxide–polymer NCs with controlled particle size, material morphology, and other properties can be achieved. Among the most promising routes to producing them in this way is the intermatrix synthesis (IMS) technique, the methodology of which is based on carrying out the following two sequential steps [13, 22, 23]:

1. Introduction of the MNP or MONP precursors into the polymer
2. Their reduction to zerovalent state (in case of MNPs) or precipitation/coprecipitation (in case of MONPs) inside the polymer matrix [24]

The first step can be carried out in the case of functionalized polymers by loading their functional groups with the desired metal ions or metal complex precursors of the NPs. The second step is carried out by using appropriate reducing (in the case of IMS of MNPs) or precipitating agents (in case of IMS of MONPs), including $NaBH_4$ or NaOH, respectively. In Section 8.1.3, a more detailed description of the synthetic methodology is reported.

It is noteworthy that although almost all publications in the field of IMS of various inorganic NPs and related areas of nanoscience and nanotechnology are dated between the end of twentieth and beginning of the twenty-first centuries [25, 26], one can also find the examples of the use of this technique both in human history and in nature [27].

8.1.2 Intermatrix Synthesis Examples from History and Nature

Intermatrix synthesis was essentially the first method used by humans for centuries in the production of various NC materials including MNP-containing glasses and ceramics, in which MNPs played the roles of very stable decorative pigments and dyes [28, 29], with the Lycurgus cup being the most cited example [30] even if similar materials can be found in Egyptian [31], Chinese [32], Celtic [33], or Vietnamese [34] traditional craftworks.

This technique was also used since the Greco-Roman period for dying wool and human hair (made of keratin and natural biopolymers), which worked by forming lead sulfide NPs in blonde hair and wool (Fig. 8.1) [35]. This example can be considered as the first known application of the IMS technique for the production of polymer–inorganic particle NCs [36]. The ancient recipe is based on immersing the blonde

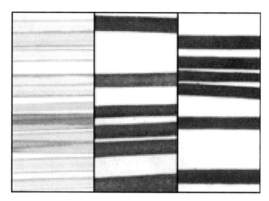

Figure 8.1. Light photos show blonde hair becoming increasingly darker after multiple applications of hair dye recreated from the ancient Greco-Roman recipe. Image adapted with permission from [36]. Copyright (2006) American Chemical Society. (*See insert for color representation of the figure.*)

Figure 8.2. Transmission electron microscopy image of magnetotactic bacteria containing magnetite NP chains, magnetosomes. Reproduced from *FEMS Microbiology Ecology*, 52 (2), 185–195. Image adapted with permission from [36]. Copyright (2006) American Chemical Society.

human or wool hair in an aqueous solution of two reagents: $Ca(OH)_2$ and PbO. The solution of lead oxide provides a source of Pb^{2+} ions. Sulfur-containing natural amino acids (cystine, cysteine, and methionine) serve as a source of sulfide ions needed for the formation of galena (PbS) NPs. Despite the structural complexity of hair, PbS NPs easily crystallize and are organized inside this biopolymer, which serves as a sort of a nanoreactor [37].

On the other hand, the intracellular synthesis of magnetic inorganic NPs (inside the natural polymers) is also known to be used by a number of microorganisms [38, 39] for self-orientation in the magnetic field. The magnetotactic bacteria (see Fig. 8.2a) discovered in 1975 by Richard P. Blakemore [40, 41] represent the best example of microorganisms using this naturally existing version of IMS technique. They utilize strings of iron-containing magnetite nanocrystals (magnetosomes; see Fig. 8.2b) [42] as an aid to navigation. The very existence of these "natural" nanostructures raises questions about how they were made and why they appear to be so stable against aggregation or Ostwald ripening. Researchers have found that magnetosome vesicles appear to be true membrane-bounded organelles controlling magnetite formation in a temporally and spatially coordinated fashion [43]. This sort of confinement can be understood as an example of IMS.

The first communication about the IMS of MNPs in functional polymers (concretely amine exchange granulated polymeric material) dates back to 1949, in which Mills and Dickinson have described the preparation of an anion-exchange resin containing Cu MNPs ("colloidal copper") and the use of this polymer–metal NC to remove oxygen from water based on its interaction with Cu MNPs [44]. Since then, many studies about the modification of ion-exchange resins with MNPs (mainly Cu MNPs) resulted in the development of a new class of NC ion-exchange materials combining the ion-exchange procedure (due to the presence of functional groups in the matrix) and the redox properties (due to the presence of "colloidal metal" or MNPs inside the matrix). They are also known as redoxites and electron exchangers [45–47]. The preparation of such materials was based on the use of the IMS technique and involved the two aforementioned consecutive stages. The redoxites have found wide application in the complex water treatment processes at power stations for the removal of hardness ions (by ion exchange) and oxygen (by redox reactions with MNPs). However, essentially no information about the sizes and structures of MNPs synthesized in redoxite matrices and the features of their distribution inside polymers can be found in the literature.

8.1.3 Development of IMS Methodology and Coprecipitation Process of Environmentally Friendly MNPs and MONPs for Catalysis and Water Treatment Applications

As previously introduced, one of the most promising routes to producing the polymer-stabilized MNPs (PSMNPs) and NCs based on these is IMS [13, 17]. The IMS procedure involves two straightforward steps: (i) the loading of the functional groups of the polymer with metal ions, followed by (ii) their either chemical reduction or coprecipitation inside the matrix. The synthetic procedure and, consequently, the properties of the final NC will be determined by several parameters, including polymer matrix type and porosity, nature of functional groups, and metal reduction conditions. For example, the functional groups of the polymer, which may be cationic or anionic, determine the choice of NP precursor and the sequence of IMS stages. As we reported before, IMS has been observed in several examples in nature, where NPs are shown to be distributed along the polymer. However, the main goal of this study is to achieve the best NP distribution for the desired applications, which means a surface NP distribution and can be achieved by utilizing the Donnan exclusion effect [48]. The Donnan effect-driven intermatrix synthesis (DEDIMS) method is based on the inability of co-ions to deeply penetrate inside the polymer when the sign of their charge is the same as that of the polymer functionality. For example, if the polymeric matrix bears negative charges due to its functional groups, then anions cannot deeply penetrate inside the matrix due to electrostatic repulsion. The action of this driving force results in the formation of NPs (in our case, both MNPs and MONPs) mainly near the surface of the polymer allowing maximal accessibility for the reagents or substances to be treated through surface contact, including bacteria to be eliminated [49–51].

As a result of this, selection of an appropriate reducing agent is of great importance. Thus, ionic reducing agents including trisodium citrate, $NaBH_4$, ascorbic acid, and hydrazine have been widely studied. Of the conventional methods for producing silver nanomaterials (NMs), the borohydride method is the most common. The main reasons are the relatively high reactivity of sodium borohydride, its moderate toxicity [52], and greater lab safety when compared to other reducing agents or physical methods. Moreover, the effect of the precipitating agent (e.g., NaOH or NH_4OH) to the MONP formation is also important. The size and shape of the NPs can be tailored with relative success by adjusting pH, ionic strength, temperature, etc.

Despite much work on IMS in our research group being focused on the use of cationic exchangers' matrices, recently, much attention has been focused on amine functionalization since amines are well known to stabilize the NPs against aggregation without disturbing the desired properties [53, 54]. Thus, this communication also reports the further development of the DEDIMS technique by extending it to polymers whose matrices bear positively charged functional groups (anionic exchange polymers or resins). The version of this methodology, which is applied in this case, can be considered as the "symmetrical reflection" of the previously developed DEDIMS method developed by cationic exchangers [55] since the first step is the loading of the functional groups of the polymer with reducing agent anions or with a highly negatively charged substance followed by the treatment of the resin with the ion salt solution precursor of the MNPs. The last stage resulted in the formation of polymer–metal NCs containing PSMNPs (Fig. 8.3).

With the goal of producing MONPs, the coupling of the coprecipitation method (ideal for magnetite NP synthesis) to the IMS technique using the Donnan effect is reported in this study. In general, the coprecipitation method involves the mixture of the NP precursor's salt solutions followed by the addition of a precipitating agent. In this process, two stages are involved. First, a fast nucleation process occurs when the concentration of the species reaches critical supersaturation (when a size control of monodispersed particles must normally be performed); then, a slow growth of the nuclei by diffusion of the solutes to the surface of the crystal takes place.

8.1.4 Magnetic NPs

Magnetic NPs are of great interest for researchers from a wide range of disciplines, including catalysis, biotechnology/biomedicine, and environmental science and technology. It has been reported that magnetic NP suspensions have a great variety of biomedical applications as MRI contrast agents and drug delivery systems [56]. Magnetic NMs for heterogeneous catalysis have also attracted much interest because of their large surface area, high activity, and recyclability [57–59], as well as because they can be conveniently recovered for reuse in sequential catalytic cycles by using an external magnetic

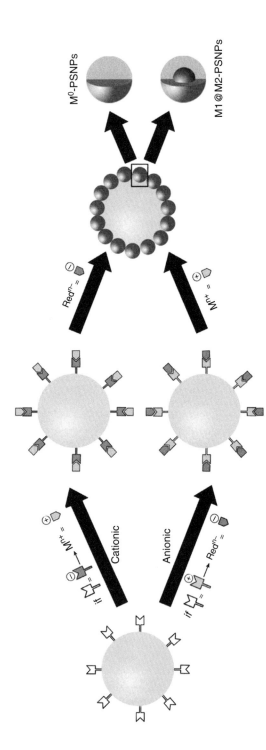

Figure 8.3. Schematic diagram of IMS steps for synthesis of NPs in granulated polymer matrix. (*See insert for color representation of the figure.*)

field [60, 61]. Up to now, several types of magnetic materials have been used including Co or Fe oxides including magnetite, hematite, maghemite, and wüstite [62, 63]. Moreover, for many practical applications, it is necessary to develop strategies to chemically stabilize the magnetic NPs against coalescence and degradation. One technique for doing this is to synthesize polymeric materials containing magnetic NPs [64, 65]. For instance, Lim *et al.* [61] reported the coating of oleic acid-stabilized superparamagnetic Fe_3O_4 NPs with a commercially available polymer.

Additionally, in our previous reports [48, 66] we have shown the magnetic properties and the catalytic activity of Pd@Co NPs stabilized in different types of ion-exchange polymeric materials by testing the magnetic recovery for repetitive catalytic cycles in cross coupling Suzuki reaction.

One of the synthetic advantages of the IMS method allows the preparation of various NP architectures. For example, core–shell MNPs can be prepared by coating the monometallic MNPs obtained after the first cycle (core) with the functional metal shell. The properties of the final NC are determined by the properties of both metals as described previously. For catalytic purposes, the surface of magnetic NPs is often chemically functionalized because of the poor catalytic properties of the bare Fe oxides or other catalytic materials [67–69]. A few nanometer-sized transition metals already displayed useful properties (e.g., Au, Pt, Pd, Ag) but do not possess magnetic properties [70, 71]. Hence, adopting a core–shell structure, NPs with a magnetic core and a functional shell is a win–win strategy [17, 72–74].

8.1.5 Nanoparticle Toxicity and Environmental Safety of Polymer–Metal NCs

A current critical issue concerning NPs is their environmental and health safety risks, sometimes referred to as nanotoxicity [75–78]. The environmental safety of NC materials (which contain various NMs) is one of the hottest topics of modern nanotechnology within the last few years. The main concerns dealing with the rapid development and commercialization of various NCs are associated with (i) the approved higher toxicity of many NMs in comparison with their bulk counterparts, (ii) the absence of the adequate analytical techniques for detection of NMs in the environment, and (iii) the absence of the legislation normative for permitted levels of various NMs in water and air. Thus, prevention of NP escape to the environment is the best approach that can be considered at the moment. In this sense, the embedding of NPs into organic or inorganic matrices and the use of magnetic NPs for their design reduce their mobility and prevent their appearance in the environment since they can be easily entrapped and recovered in the case of their leakage from the NC by using simple magnetic traps [17, 64]. In our case, by using the IMS method, PSMNPs are strongly captured inside the polymer matrix. Moreover, in the case of bimetallic magnetic core–shell MNPs, one can also increase the NC safety since MNPs leached from the polymer matrix can be easily captured by the magnetic traps that permit either complete prevention of any post-contamination of the treated medium or recovery and recycling of the MNPs.

This work focuses on the synthesis of superparamagnetic core NPs, which are represented by iron oxide NPs and which have traditionally been synthesized by coprecipitating Fe^{2+} and Fe^{3+} metal ions. To produce iron oxide NPs, a wide variety of factors can be adjusted to control the size, magnetic characteristics, or surface properties [61]. A number of factors related to the nature of the polymer support also have a significant effect on the iron oxides' synthesis: functional group properties including pK_a as well as polymer macrostructure including porosity that affects the swelling capacity, among other factors.

In this work, the influence of those parameters, based on the polymer, magnetic properties, and NP distribution, has been studied in a basic coprecipitation process. Thus, the use of both sulfonated and amine functionalized polymers, mainly granulated resins for IMS of NPs of various compositions, is investigated [55]. The IMS and characterization of magnetite/silver core–shell NP is reported. The synthesis was performed by a two-step procedure that involves the precipitation of magnetite NPs from coordination between ferric and ferrous ions with the polymeric functional groups and subsequent reduction of silver ions onto the magnetite NP surface and within the matrix. This preparation method allows the preparation of highly loaded polymer NCs with magnetite/silver core–shell NPs. This is related to our previous works based on IMS of Ag@Co and Pd@Co NCs stabilized on different ion-exchange polymeric materials including membranes, nonwoven fibers, and granulated resins [48, 49].

8.2 SYNTHESIS OF Ag@Co AND Pd@Co NCs BY DEDIMS WITH ENHANCED DISTRIBUTION OF MNPS FOR CATALYTIC AND BIOCIDE APPLICATIONS

Ag@Co and Pd@Co NC synthesis was carried out as reported elsewhere [48, 49]. Our previous work described the synthesis of core–shell MNPs consisting of a Co magnetic core covered by a functional metal using IMS in ion-exchange polymeric matrices [66]. Two different core–shell NP structures were obtained depending on the functional metal: Ag@Co NPs with bactericidal activity and Pd@Co NPs with catalytic capacity. Therefore, using this protocol, bifunctional NPs presenting both magnetic and functional properties on several types of ion-exchange polymeric matrices (membranes, nonwoven fibers, and granulated supports) have been obtained. Additionally, with the final application of interest in mind for these NCs (catalytic or bactericide activity) and with the goal of obtaining an appropriate distribution of NPs in the polymer, IMS methodology with Donnan effect has been adapted. Ensuring a good NP distribution enables enhanced access of the substrates to the functional part of the NC.

The IMS procedure of preparing PSMNPs involves two relatively straightforward steps: (i) the loading of the functional groups of the polymer with metal ions (e.g., Co^{2+}), followed by (ii) their chemical reduction inside the matrix resulting in the formation of monometallic MNPs. In most cases, $NaBH_4$ has been used as the reducing agent due to its efficacy for the IMS of MNPs of various compositions and structures [13]. These two steps may be described by the following equations for a cationic exchanger, where R is any organic or alkyl functionality:

$$2\text{R-SO}_3^-\text{Na}^+ + \text{Co}^{2+} \rightarrow \left(\text{R-SO}_3^-\right)_2 \text{Co}^{2+} + 2\text{Na}^+ \qquad (8.1)$$

$$\left(\text{R-SO}_3^-\right)_2 \text{Co}^{2+} + 2\text{NaBH}_4 + 6\text{H}_2\text{O} \rightarrow 2\text{R-SO}_3^-\text{Na}^+ + 7\text{H}_2 \\ + 2\text{B(OH)}_3 + \text{Co}^0 (\text{Co MNPs})$$

(8.2)

Afterwards, with the goal of coating the magnetic core with a functional metal (e.g., Ag$^+$), IMS was applied, following steps (i) and (ii) as described earlier:

$$\text{R-SO}_3^-\text{Na}^+ + \text{Ag}^+ + \text{Co}^0 \rightarrow \text{R-SO}_3^-\text{Ag}^+ + \text{Na}^+ + \text{Co}^0$$

(8.3)

$$\text{R-SO}_3^-\text{Ag}^+ + \text{Co}^0 + \text{NaBH}_4 + 3\text{H}_2\text{O} \rightarrow \text{R-SO}_3^-\text{Na}^+ + 7/2\text{H}_2 \\ + \text{B(OH)}_3 + \text{Ag@Co}$$

(8.4)

The synthesis of Ag@Co NPs has been achieved in a large number of functional polymer materials obtaining core–shell structure [49] as well as showing high bactericide properties [79].

Figure 8.4 shows the Co NP distribution on a cross section of sulfonated polymer (granulated material) sample and the corresponding magnetic behavior curve. A surface NP distribution as well as the superparamagnetic properties of the NC is observed.

Figure 8.5 shows the transmission electron microscopy (TEM) image of Pd@Co NCs in fibrous material. In this case one can clearly see the presence of MNPs with core–shell structures where the light core corresponding to Co metal shell is surrounded by a darker Pd shell [48]. The difference in contrast of core and the shell is associated with the difference of densities of Co and Pd metals (8900 and 12,023 kg·m^{-3}, respectively). Moreover, to demonstrate the catalytic utility of the composite, the Pd@Co NCs in fibrous material were used as catalyst for Suzuki reaction as previously introduced. Figure 8.5b shows the Suzuki reaction scheme at the experimental conditions when the Pd@Co carboxylated fibrous material is used as a catalyst. The product yields for four runs (or cycles after recovery by magnetic separation) are also shown, demonstrating its recyclability.

The DEDIMS protocol was also successfully applied to anion-exchange polymers containing positively charged functional groups. It is described in Equations (8.5) and (8.6), where R-NH$_3^+$ is the anionic polymeric exchanger, X are the anions that exchange with the functional groups of the polymer, R$_{ed}^{n-}$ corresponds to an anionic reducing agent, and M^{n+} is the metal that is reduced on the polymer, where n is any positive integer:

$$\text{R-NH}_3\text{X}_1 + \text{R}_{ed}^{n-} \rightarrow \left(\text{R-NH}_3^+\right)_n \text{R}_{red}^{n-} + n\text{X}_1^-$$

(8.5)

$$\left(\text{R-NH}_3^+\right)_n \text{R}_{red}^{n-} + \text{M}^{n+}\text{X}_2^{n-} \rightarrow n\text{R-NH}_3\text{X}_2 + \text{M}^0 + \text{R}_{ox}$$

(8.6)

The IMS of PSMNPs using anionic polymers was carried out by loading the functional groups of the polymer with reducing agent anions using aqueous NaBH$_4$ solution, followed by the treatment of the resin with the ion salt solution precursor of the MNPs. The last stage resulted in the formation of polymer-stabilized metal nanocomposites (PSMNCs). A number of interesting results were obtained by the synthesis of monometallic Ag and Pd NPs on ion-exchange granulated polymers containing quaternary ammonium functional groups [55].

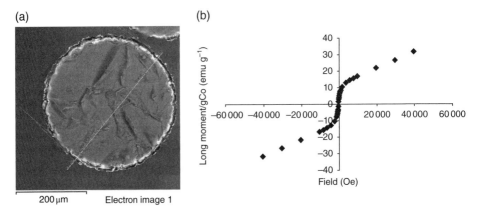

Figure 8.4. (a) Scanning electron microscope image of cross section of granulated Co MNP–sulfonated polymer NC; red line across bead diameter corresponds to EDS spectrum, which shows Co MNP distribution and (b) magnetization curve of NC sample. (*See insert for color representation of the figure.*)

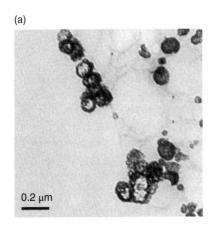

(b)

Pd1@Co-FIBAN K-4
(2 mol% Pd)
K$_2$CO$_3$

DMF/H$_2$O 4:1
80°C, 18 h

%Yield	1st	2nd	3th	4th
Pd1@Co	100	92.9	36.7	15.5

Figure 8.5. Characterization of Pd@Co MNPs stabilized on carboxylated fibrous materials by (a) TEM image and (b) Suzuki reaction catalysis after four runs. Percentage of GC yields is corrected to internal standard n-C$_{11}$H$_{24}$.

(a) (b)

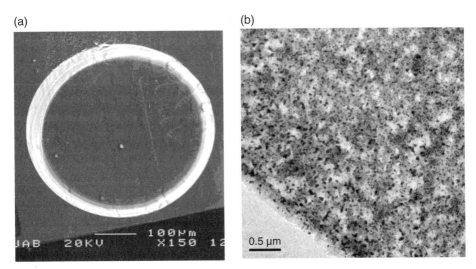

Figure 8.6. (a) Scanning electron microscopy image of cross section of Ag–520E NC prepared by using 0.1 M $Na_2S_2O_4$ solution for metal reduction and (b) TEM image of boundary part of same cross section of the sample. $AgNO_3$ 0.1 M was used at first stage of IMS of Ag MNPs (see schematic diagram in Figure 8.3).

The importance of the metal reduction conditions, namely, the type of the reducing agent applied, in the reactions represented in Figure 8.3 must be taken into account. Indeed, the monocharged borohydride anions can penetrate deeper into the polymer matrix bearing a charge of the same sign as a doubly charged reducing agent including one containing dithionite anions due to the Donnan exclusion effect. Examination of the features and distribution of MNPs obtained by IMS coupled with Donnan effect inside the polymer helps us to explain this [66]. The use of $NaBH_4$ as a reducing agent was described previously (Fig. 8.4). The use of $S_2O_4^{2-}$ as a reducing agent is shown in Figure 8.6, which corresponds to Ag NPs distribution in the macroporous anionic exchanger polymer (R-NH_3^+) prepared using $S_2O_4^{2-}$ as reducing agent at 0.1 M concentration.

Ag NPs are clearly well distributed on the surface of the polymer, as shown by a high-intensity circle around the cross section of the polymer bead (Fig. 8.6a). The presence of nanometer-sized particles is also observed (Fig. 8.6b).

When preparing core–shell NPs, it is important to know the nature of the oxidative states of the metals containing polymer–metal NCs in order to determine whether the desired composite has been synthesized. Thus, X-ray absorption near-edge structure (XANES) technique was used. X-ray absorption near-edge structure spectra supply information related to atomic organization and chemical bonding by comparison (linear combination) with standards. Samples tested were Ag, Co, Pd, Ag@Co, or Pd@Co PSNPs in different matrices. The samples were tested against metal standards including metal foils (zero state) and salts. The standards were measured in order to calibrate the energies of the edge positions for Ag, Co, and Pd in different environments. The results of representative samples are illustrated in Figure 8.7. The analysis procedure by the Co

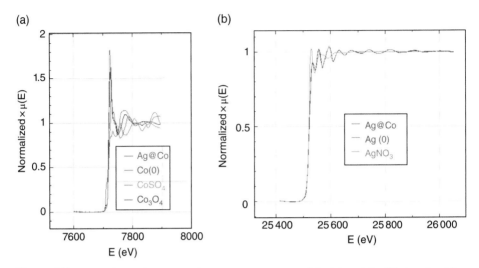

Figure 8.7. X-ray absorption near-edge structure spectra of Co (a) and Ag (b) signals of the metal Ag@Co carboxylated resin sample (blue line in both graphics) in comparison with Co and Ag standards. (a) Co^0 (red), CoSO$_4$·7H$_2$O (green), Co$_3$O$_4$ (violet), and (b) Ag^0 (red) and AgNO$_3$ (green) are standards. (*See insert for color representation of the figure.*)

K-edge and Ag L-edge X-ray absorption spectra (XAS) was recorded on the BM25A beam line of the ESRF synchrotron source (Grenoble, France) with a ring current of 120–170 mA at 2.5 GeV. The data were collected in transmission mode with the nitrogen (85%) and argon (15%) gas-filled ionization chambers as detectors at room temperature. Energy calibrations were carried out with the Co and Ag metal foils, assigning the first inflection point to 7709 and 25,514 eV, respectively. X-ray absorption near-edge structure data were analyzed and modelled using the IFEFFIT-based suite of programs, ATHENA and ARTEMIS, developed by Ravel and Newville [7].

In Figure 8.7, the edge energy for different Ag and Co standards can be observed. The spectrum lines for each element (Co and Ag) present in Ag@Co carboxylated polymer are clearly shown. These corresponded to the average of all measurements done for the corresponding sample. By comparison with the spectra of the standards, we can determine the oxidation state of the corresponding metal in each sample.

Figure 8.7a shows that the merge spectrum for Co was between Co$_3$O$_4$ and CoSO$_4$, indicating that metallic Co was partially oxidized to either +2 or +3 oxidation states. X-ray absorption near-edge structure spectra were consistent with the core magnetic phase being primarily Co, Co^{2+}, and Co^{3+}. The key issues here are the chemical states of the core materials and whether the oxide forms during or after the synthesis process.

Conversely, Figure 8.7b suggests that most of the Ag in the NP structure remained as Ag^0 (shown by the perfect match between Ag (0) and Ag@Co spectrum lines), which was optimal for their application as bactericide material. Monometallic Co NP NC spectra (data not shown in this report) displayed the same tendency as Figure 8.7a, which may mean that Co oxidation may have occurred during the Co loading, after the reduction step or due to the silver loading [80].

Moreover, oxidation of the Co of these Co core NPs after the shell coating indicates possible Co release into the medium. Additionally, due to their synthetically complex routes (e.g., for the synthesis of Co NPs on anionic resins), the development of another nontoxic and stable supermagnetic core was necessary. Thus, the following sections are based on the synthesis of iron oxide-based polymer-stabilized NCs and their characterization.

8.3 GENERAL DESCRIPTION FOR THE COPRECIPITATION OF MFe$_2$O$_4$ NPs (WHERE M=M^{2+} AND CAN BE Fe^{2+}, Mn^{2+}, OR Co^{2+}) BY THE IMS PROCEDURE COUPLED WITH COPRECIPITATION

The DEDIMS method coupled with the coprecipitation technique can be applied to obtain iron oxide NPs including Fe$_3$O$_4$, MnFe$_2$O$_4$, and CoFe$_2$O$_4$ within a polymer matrix [81]. Moreover, the low level of toxicity and the superparamagnetic properties of many ferrous oxide NPs are very convenient for biological applications [82].

The iron oxide NPs can be synthesized by using the traditional method, based on the coprecipitation of transition metal ions, including Fe^{2+} and Fe^{3+}, in aqueous solutions. To produce monodisperse iron oxide NPs, the size and shape, magnetic characteristics, or surface properties of the NPs can be tailored by adjusting pH, ionic strength, temperature, nature of the salts, or the reagent concentration ratio [61]. Factors related to the nature of the polymer support also have a significant effect on the iron oxide synthesis: functional group properties, including pK$_a$ and polymer macrostructure including porosity that affects the swelling capacity, among others. For example, the addition of organic anions (carboxylates, including citric, gluconic, or oleic acid) during the formation of magnetite can help to control the size of the NPs. The influence of those parameters on magnetic properties and NP distribution has been studied in a basic coprecipitation process [82].

Therefore, a number of synthetic procedures were investigated to coat the iron oxide cores with silver shell to obtain Ag@Fe$_3$O$_4$ NPs with bactericidal activity. The chemical reaction of Fe$_3$O$_4$ formation may be written as follows:

$$Fe^{2+} + 2Fe^{3+} + 8OH^- \rightarrow Fe_3O_4 + 4H_2O \tag{8.7}$$

Complete precipitation of Fe$_3$O$_4$ should be expected at a pH between 8 and 14, with a stoichiometric ratio of 2:1 (Fe^{3+}/Fe^{2+}) in a nonoxidizing environment. In general, under an argon atmosphere, a concentrated aqueous solution of sodium hydroxide was added to the Fe^{3+}/Fe^{2+} mixture, immediately turning it to a dark-brown/black precipitate. The solution was stirred and heated, which resulted in the formation of a brown colloidal solution of ferrite NPs [53]. The effect of Fe^{3+}/Fe^{2+} concentration on the yield and the polydispersity of the NPs was carefully studied. The specific synthetic conditions will be detailed in each section.

The matrices, based on polystyrene cross-linked with divinylbenzene (DVB), used as a polymeric support in our study are detailed in Table 8.1. Carboxylic ion exchangers are based on carboxylated polyacrylic acid cross-linked with DVB (macroporous type), whereas the sulfonic ones are based on sulfonated polypropylene fiber copolymerized

TABLE 8.1. Main Characteristics of Polymers Used for the Preparation of Polymer–Metal NCs in This Work, Including Functional Group and IEC

Name	Functional group	Classification	IEC (meq g^{-1})
SPEEK	R-SO$_3^-$	Nonporous membrane	2.6
C100E	R-SO$_3^-$	Gel resin	2.3
SST80	R-SO$_3^-$	Gel resin	4.2
A520E	R-NH$_3^+$	Macroporous resin	1.4

with styrene and DVB (gel type). Their ion-exchange capacities (IEC) as well as some of their main characteristics are also indicated in Table 8.1. Note that the main results described in this chapter have been obtained with the NCs synthesized based on sulfonated polymers: sulfonated polyether ether ketone (SPEEK) and granulated polystyrene resins, C100E and SST80.

The main difference in the properties of C100E and SST80 polymers is based on the distribution of functional groups (sulfonic) in the polymer matrix. In the first case, groups are homogeneously distributed throughout the matrix, while in the second one, the functional groups are distributed on the surface (outer shell) of the polymeric bead as its preparation is based on the so-called shallow shell technology [83]. Non-cross-linked SPEEK polymer was used as a model polymeric matrix due to the similarity of its functionality to other sulfonated polymers as it has been used in our previous works [84, 85].

Although an intensive interest has been devoted within the last decade to the use of polymeric materials as supports for catalysts [86], reagents, or even enzymes, more research is still required to elucidate the influence of the structural parameters of the support (including porous texture, cross-linking degree, location, and distribution of the functional groups, to name a few) onto the properties of the final material in its final applications (including catalytic activity or selectivity) [87]. The ion-exchange resins are usually classified by the structural features of their polymeric matrices as "gel type" or "macroporous type," which gives an additional parameter to determine their suitability in various practical applications. The term "macroporous" is usually applied to the resin, which has permanent porosity and very large surface area even in the dry state. On the contrary, the gel-type resins in the dry state are characterized by very low porosity and a far lower surface area, which usually does not exceed 5 m$^2 \cdot$g^{-1} (N$_2$ sorption, BET) [88].

8.3.1 Intermatrix Synthesis of Fe$_3$O$_4$ NPs in Sulfonated Polymers SPEEK, C100E, and STT80

The IMS of Fe$_3$O$_4$ NPs was carried out by adapting the general synthetic procedure described elsewhere [89] to the interpolymer conditions. A concentrated aqueous solution of sodium hydroxide was added into a mixture of iron salts with Fe^{2+}/Fe^{3+} molar ratio of 1:2 that was mixed with an exact amount of ion-exchange polymer under stirring and an Ar atmosphere. Immediately, the polymer became black brown in color. The reaction mixture was intensively stirred, which resulted in the formation of magnetite particles on the polymeric material. The magnetic beads of the polymer–Fe$_3$O$_4$ NC were washed and dried in an oven at 60 °C. The IMS of Fe_3O_4 NPs in sulfonated polymers can be described by the following reaction scheme:

$$5\left(R\text{-}SO_3^-Na^+\right) + Fe^{2+} + 2Fe^{3+} \rightarrow \left(R\text{-}SO_3^-\right)_2\left(Fe^{2+}\right) + \left(R\text{-}SO_3^-\right)_3\left(Fe^{3+}\right) + 5Na^+ \quad (8.8)$$

$$\left(R\text{-}SO^{3-}\right)_2\left(Fe^{2+}\right) + 2\left(R\text{-}SO^{3-}\right)_3\left(Fe^{3+}\right) + 8NaOH^- \rightarrow 8R\text{-}SO_3^-Na^+ \\ + Fe_3O_4 + 4H_2O \quad (8.9)$$

8.3.2 Synthesis of Fe_3O_4 NPs in Quaternary Ammonium Polymer, A520E

The synthesis of Fe_3O_4 NPs in A520E polymer is a new challenge in terms of NP synthesis by IMS. Recently, much attention has been focused on the use of NP-stabilizing agents with amine functionalization since amines are well known to stabilize NPs against aggregation without modifying their desired properties. Also, in the classical preparation of colloids of precious metals, including gold, palladium, and silver, sodium citrate is used as both the reducing agent and stabilizer of the resulting particles. Citrate is an efficient stabilizer but, depending on its concentration, it has a strong influence on the size and morphology of the silver particles formed [53, 90]. Due to the anionic nature of the polymer, the "symmetrical reflection" of the traditionally used IMS procedure should be used in this case by taking into account the positive charge of the functional amino groups (see Fig. 8.3). Thus, a highly charged stabilizer was needed (such as sodium citrate) prior to the ion-exchange reaction with iron ions for the formation of Fe_3O_4 NPs. Afterwards, the NC was loaded with the iron salt mixtures followed by the addition of the NaOH, as already described.

In general, for all the synthetic procedures, different basic media including NaOH, NH_4OH, and TMAOH may be used to bring the synthesis to completion. The difference between them was based on the speed of the precipitation reaction that modified the NP synthesis. In this sense, higher pH and ionic strength values result in the formation of smaller particles with a wider size distribution. These parameters have been shown to determine the chemical composition of the crystal surface and consequently the electrostatic surface charge of the particles [91], as will be described later.

Also, in some cases, temperature played an important role both during and after the magnetic NP synthesis. For instance, in the synthesis of Fe_3O_4 in A520E polymer, a temperature higher than 80 °C was essential to obtain the final NP structure. On the other hand, one also has to take into account that magnetite can discompose to hematite under high temperature conditions.

8.3.3 Intermatrix Synthesis of Polymer–Ag@Fe_3O_4 NCs

Despite the significant volume of research into the synthesis of magnetic NPs of different compositions and sizes, their long-term stability in suspension without aggregation and precipitation still remains an unsolved problem. In fact, upon formation, precipitated

TABLE 8.2. Synthetic Conditions for Preparation of Fe_3O_4 and $Ag@Fe_3O_4$ Samples

Matrix	Sample containing Fe_3O_4	Synthetic conditions	Sample containing $Ag@Fe_3O_4$	Synthetic conditions
SPEEK	S	Coprecipitation		
C100E	C	Coprecipitation	Ag-C	IMS reduction
	CH	Coprecipitation, low conc.	Ag-CH	IMS reduction, Ag 0.01M
SST80	SF	Coprecipitation	Ag-SF	IMS reduction
	SH	Coprecipitation, low conc.	Ag-SH	IMS reduction, Ag 0.01M
A520E	A	Coprecipitation, with Na citrate, 2h	Ag-A	IMS reduction
Non-polymeric matrix	PF	Coprecipitation		

ferrite particles tend to aggregate quickly. The most popular method to guard against this difficulty is surface modification. The use of a surfactant or polymer is the most common technique to stabilize NPs in suspension. However, polymer stabilizers are not ideal, as they do not provide a robust impenetrable shell; additionally, their suscepti- bility to degradation at high temperatures limits their use. An alternative NP coating material can be a metal. Recent research has shown that NPs can be coated with a layer of metal, which serves to both protect and stabilize them. For example, Ban and coworkers have coated Fe NPs with a thin layer of Au to reveal NPs with a core–shell structure [92].

As it has been recently reported [93], ion-exchange functional groups presented a remarkable affinity to coordinate transition metal ions including Fe(II), Co(II), Cu(II), Ni(II), and Pb(II). Moreover, polymers presented relative permeability to the diffusion of silver ions, suggesting the possibility of synthesizing magnetite/silver core–shell NPs ($Ag@Fe_3O_4$) using polymeric materials as stabilization media.

Following the IMS procedure described for MNPs, the silver covering on Fe_3O_4 core NPs was performed by a simple $AgNO_3$ loading and $NaBH_4$ reduction step. The same protocol was used for all the composites.

However, as mentioned previously, the nature and concentration of the reducing agent are known to significantly influence the conditions of NP synthesis and also the amount of NPs formed and their distribution inside the stabilizing matrix [66, 94, 95].

It is known that mild experimental conditions promote the reduction of Ag(I) ions adsorbed onto Fe_3O_4 particles. For instance, reduction by using a mild reducing agent was employed in order to ensure a controlled shell growth of silver onto Fe_3O_4 particles and to avoid the formation of new silver nuclei outside the polymer matrix [96]. The general IMS scheme in this case can be described by the following equations:

$$R\text{-}SO_3^-Na^+ + Fe_3O_4 + Ag^+ \rightarrow R\text{-}SO_3^-Ag^+ + Na^+ + Fe_3O_4 \qquad (8.10)$$

$$R\text{-}SO_3^-Ag^+ + Fe_3O_4 + NaBH_4 + 3H_2O \rightarrow R\text{-}SO_3^-Na^+ + 3/2H_2$$
$$+ B(OH)_3 + Ag@Fe_3O_4 \qquad (8.11)$$

Table 8.2 shows the NC samples prepared using sulfonated polymeric matrices and the NMs synthesized by liquid-suspension method (without matrix). The column entitled synthetic conditions describes the reaction used and any parameter variations. For example, coprecipitation refers to the coprecipitation method previously described; coprecipitation low conc. refers to the same reaction procedure but with half the concentration of iron salts. Ag samples were synthesized by using a 0.01 M of $AgNO_3$.

8.4 CHARACTERIZATION OF POLYMER–METAL OR POLYMER–METAL OXIDE NCs

As it has been previously reported [84, 85], the IMS technique makes it possible to prepare both mono-component and bicomponent NPs (metal or metal oxide based) in both anion- and cation-exchange matrices. In all cases, the polymer–metal NCs demonstrate quite strong magnetic properties. A number of parameters including NP size and spatial distribution were studied by examining sample cross sections by scanning electron microscopy (SEM) and TEM. In all cases, Ag@Co and Pd@Co NPs synthesized in different types of polymer matrices were distributed heterogeneously inside polymer matrices with higher concentrations found on the surface of the NCs. This distribution is favorable for the applications under study, both water disinfection and catalytic applications, respectively.

For the new $Ag@Fe_3O_4$ NC materials described in this report, additional characterization techniques including thermogravimetric analysis (TGA) and X-ray diffraction (XRD) were used to further investigate the composition of the resulting materials.

Inductively coupled plasma–atomic emission spectrometry (ICP–AES) technique was used to determine the amount of each metal present in a PSMNP sample. Experimentally, NC fragments of known weight were digested with concentrated HNO_3, diluted and filtered using Millipore filters, and analyzed. The metal content was determined by ICP–AES using an Iris Intrepid II XSP spectrometer (Thermo Electron Co.) and ICP–MS (Agilent 7500). The metal content of all the elements of the $Ag@Fe_3O_4$ and Fe_3O_4 NC samples is detailed in Table 8.3 in terms of $mg_M\ g_{NC}^{-1}$ and of $mmol_M\ meq_R^{-1}$ (where M refers to the metallic element of the NPs and NC to the overall nanocomposite and meq_R to the milliequivalents of the functional groups of the polymer). The instrumental average uncertainty of metal ions determination was in all cases lower than 2%.

In general, the amount of iron metal in the core NPs did not significantly change after coating with the Ag shell. The Ag amount in $Ag@Fe_3O_4^-$ based NCs appears to be comparable with that recently reported for polymer–$Ag@Co$ NCs [49]. There are several factors that can affect the difference in the metal content among all the types of polymers, including the IEC, porosity, and the dispersion of the functional groups.

TABLE 8.3. Metal, M, Content in Fe_3O_4 and $Ag@Fe_3O_4$ PSMNP-Based NCs, NC ($mg_M \, g_{NC}^{-1}$ and $mmol_M \, meq_R^{-1}$ meq_R = Milliequivalents of Functional Groups of Polymer)

		Metal concentration						
		$mg_M \, g_{NC}^{-1}$	$mmol_M \, meq_R^{-1}$		$mg_M \, g_{NC}^{-1}$		$mmol_M \, meq_R^{-1}$	
Matrix	Sample	Fe	Fe	Sample	Ag	Fe	Ag	Fe
SPEEK	S	42.8	0.30	Ag-S	308	39.4	1.1	0.27
C100E	C	80.3	0.63	Ag-C	478	70.0	1.9	0.55
	CH	53.4	0.42	Ag-CH	263	42.0	1.1	0.33
SST80	SF	164	0.68	Ag-SF	610	145	1.3	0.60
	SH	94.0	0.40	Ag-SH	429	69.0	0.92	0.29
				SAg	579	—	1.25	—
A520E	A	27.7	0.35	Ag-A	260	27.0	1.7	0.35

However, by taking into account the IEC (represented as meq term), it seems there is no difference on the loading capacity for the two sulfonated resins (C100E and SST80) since the mmol of metal meq^{-1} of functional group have similar values.

Thermogravimetric analysis measures changes in the sample weight as a function of temperature and/or time. Thermogravimetric analysis is used to determine polymer degradation temperatures in polymer or composite materials [97].

In this work, thermograms were obtained with a PerkinElmer Pyris 1 TGA thermogravimetric analyzer. Samples were heated to 900 °C at 10 °C min^{-1} under air atmosphere. Figure 8.8 shows the weight loss (in terms of %) versus temperature thermogram (TGA) for magnetite–C100E and silver/magnetite–C100E sulfonated NC material.

As can be seen in Figure 8.8a, the TGA curves for all three C100E samples are characterized by four weight-loss regions, which can be described as follows.

1. The weight loss between 30 °C and 400 °C can be mainly attributed to strongly adsorbed water molecules (both "free" and "bound") and surface groups from the polymer and the NPs, where applicable.

2. A significant weight loss at 450 °C for all samples. It is particularly big for the raw polymer (NP-free) in comparison with the NP-modified polymers and can be associated with the loss of the free sulfonate functionalities on the polymer.

3. The third weight loss is observed between 500 °C and 700 °C and may be attributed to the degradation of the polymer side chains. Again, the NC samples appear to be more stable than the initial polymer.

4. Finally, the weight changes at temperatures higher than 700 °C may be caused by further thermodegradation of the polymer and the phase changes of the NP material (e.g., from magnetite to maghemite). However, no dependence on the sample composition can be distinguished in this temperature range.

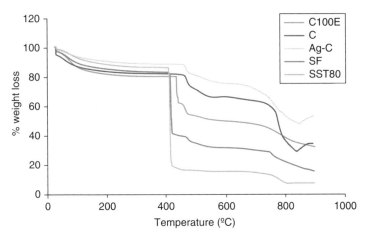

Figure 8.8. Thermogravimetric analysis curves of weight loss versus temperature for sulfonated polymer samples: C100E NP-free polymer; C sample (Fe$_3$O$_4$–C100E); Ag-C sample (Ag@Fe$_3$O$_4$–C100E); SST80 NP-free polymer and SF sample (Fe$_3$O$_4$–SST80) (see also Table 8.2). (*See insert for color representation of the figure.*)

As can be seen also in Figure 8.8, TGA curves of SST80-based samples demonstrate a very similar trend. Indeed, the NC sample appears to be more thermostable than the initial polymer, and both SST80-based samples are characterized by the same four weight-loss regions as C100E-based ones.

A general conclusion, which follows from the results shown in Figure 8.8, can be formulated as follows. All NCs showed relevant changes in TGA curves in air with respect to the neat polymer. The residue at the end of the analysis for the PSMNPs was about 10–20% of the initial weight, which was consistent with the amount of metal inorganic fraction contained in the composites. A possible explanation for the increase in weight-loss temperatures may be the formation of a more thermally stable phase, arising from a chemical mechanism. The presence of metal functionalities may lead to a partial matrix cross-linking, thus resulting in an increased thermal stability [98, 99]. Another conclusion concerns a higher thermostability of Ag@Fe$_3$O$_4$–C100E NCs in comparison with Fe$_3$O$_4$–C100E that can be assigned to the protective coating of Fe$_3$O$_4$ core NPs with a Ag shell.

X-ray diffraction technique was used to obtain the crystalline structure of the particles. In a diffraction pattern, the location of the peaks on the 2θ scale can be compared to reference peaks to allow confirmation that the desired phase of iron oxide had been prepared. Figure 8.9 shows the XRD graphs of magnetite NPs as a reference (synthesized by liquid phase method) and sample C corresponding to magnetite NPs synthesized in sulfonated C100E polymer. The samples were deposited on glass substrates. Diffraction patterns were collected on a Siemens D500 X-ray diffractometer, supplied with a Cu cathode K (alpha 1) with a wavelength of 1.54056 Å.

Figure 8.9 shows that the position and relative intensity of all diffraction peaks are in good agreement with those of the magnetite powder. The relative intensity is lower for the NC sample due to the "diluting" polymer effect.

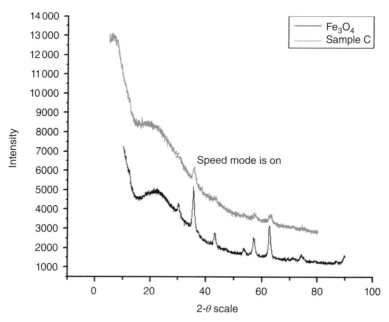

Figure 8.9. X-ray diffraction patterns of magnetite NPs: polymer-free (black line) and in sample C (red line). (*See insert for color representation of the figure.*)

8.5 SCANNING ELECTRON MICROSCOPY

Scanning electron microscopy coupled with an energy-dispersive spectrometer (EDS), Zeiss EVO MA 10 SEM, was used to characterize the polymer–NP NC material [8]. The metal concentration profiles along the cross section of the PSMNC-containing materials and the morphology of the polymer surface were also examined by using this technique. The NC samples were prepared by embedding several granules in the epoxy resin followed by cutting and cross-sectioning with a Leica EM UC6 ultramicrotome using a $35°$ diamond knife (DiATOME). When analyzing the cross section of Fe_3O_4 and Ag@ Fe_3O_4 PSMNP granules, the near-surface distribution of MNPs may be observed clearly (Fig. 8.10). Actually, most of the metal was found near the bead surface, and only few particles were detected in a deeper bead region. Indeed, EDS analysis demonstrated that all the metals were mostly found on the surface. However, in some cases, Ag structures were distributed more homogeneously along the cross section.

In order to evaluate the polymer effect on the synthesis when the same kind of NPs were synthesized under the same conditions, samples C (Fe_3O_4 in C100E matrix, not shown) and SF (Fe_3O_4 in SST80 matrix) were compared. The comparison showed a higher and deeper metal distribution in sample SF, which can be attributed to a higher concentration of functional groups in this polymer. Indeed, as it follows from the data shown in Table 8.1, SST80 polymer has an IEC value twice as high as the C100E sample.

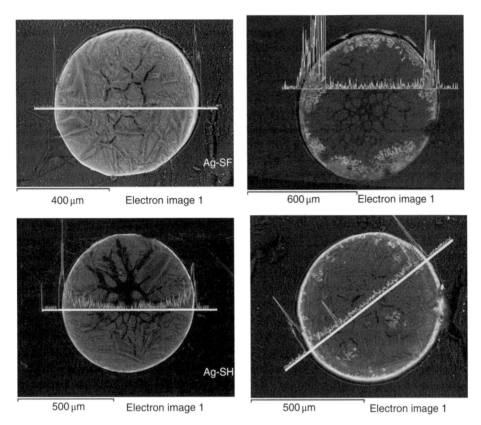

Figure 8.10. Scanning electron microscopy images of the Fe_3O_4 and $Ag@Fe_3O_4$ NC cross sections for samples SF, Ag-SF, SH, and Ag-SH (see Table 8.1). The lines show distribution of Ag (blue) and Fe (red) metal ions across particle diameters. (*See insert for color representation of the figure.*)

Therefore, the distribution of the functional groups has an important role in IMS. Taking into the account the metal concentration of both SF and SH samples, the more narrow NP size distribution was obtained for the SH sample, where the metal content is lower.

In comparison with the fibrous NC material characterized in our previous publication [79] where Co NPs were distributed quite homogeneously along the fiber cross section, the use of granulated polymers and the new IMS procedure developed permits almost a perfect matching of Ag and Fe peaks in the EDS spectra (see Fig. 8.10, Ag-SF and Ag-SH samples) that may indicate the formation of Ag shell covering Fe core NPs.

Additionally, a number of $Ag@Fe_3O_4$ NCs showed the formation of silver fractals (see Fig. 8.10, image Ag-SF and Ag-SH, and Fig. 8.11a), whose dimensions depended on the synthetic conditions and the nature of the NC (mainly for SST polymers). The formation of fractals may be caused by diffusion-limited particles aggregation inside the polymer phase acting as NP stabilizer. This phenomenon was found to be affected by the presence of NaOH in the synthetic process. The use of NaOH in the synthetic

(a)   (b)   (c)

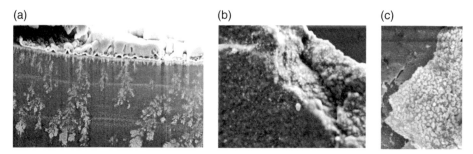

Figure 8.11. Scanning electron microscopy images of Ag@Fe$_3$O$_4$–C100E NC bead cross sections (a) obtained by using FIB technique (where Ag fractals are shown) and (b) and (c) SEM images of NC surface.

procedure acts as an accelerator and may also influence the formation of aggregation structures. The random fractals observed in this work (see also Fig. 8.11a) corresponded to the so-called Brownian tree-type fractals associated with a diffusion-limited aggregation and a reduction-limited aggregation. It is reported [100] that the higher the temperature during the reduction step, the higher the fractal dimension, which means that larger and denser particles are created. Similarly in the case shown, the samples with a higher silver fractal formation were the ones that required a higher temperature for NP formation (100 °C) as well as the SST ones (probably due to distribution of the functional groups) (Fig. 8.10). Therefore, according to the particle size, it may be assumed that most of them should be located at the external surface of the material. For instance, in PVP stabilization of Ag NPs, an increase in NaOH concentration leads to the destruction of the electrical double layer protecting NP surfaces against coalescence and, consequently, the formation of NP aggregates (fractals or dendritic structures) [101]. Also, Guodan Wei et al. [102] reported that the presence of trisodium citrate solution at high temperature in a AgNO$_3$ solution favored the formation of Ag fractals.

In our previous recent publication, we reported [55, 66] the results of TEM characterization of Pd@Co NPs stabilized in granulated polymeric matrices. However, the same procedure was not successful for the samples described in this report. In order to determine the presence of NPs in the polymeric matrix, Zeiss MERLIN FE-SEM and focused ion beam (FIB) technique was used. Figure 8.11 shows SEM images of the cross section and surface of a C100E polymeric bead containing Ag@Fe$_3$O$_4$–C100E (sample C). The SEM–FIB image of the sample shows the presence of Ag fractals in the sample. The SEM images shown in Figure 8.11b and c clearly show the presence of nanoparticulate structures on the surface, which are the Ag@Fe$_3$O$_4$ NPs.

For the characterization of the magnetic properties of the NC, a vibrating sample magnetometer (VSM) and a superconducting quantum interference device (SQUID) were used. A VSM is an instrument capable of measuring the magnetic behavior of magnetic materials. The one used in our study was based on the use of Halbach cylinder magnets up to a magnetic field of 1.0 T and operated on Faraday's Law of

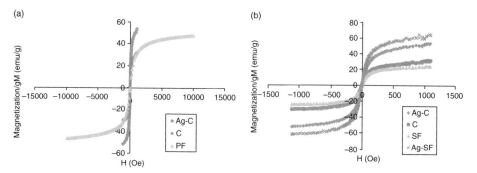

Figure 8.12. Magnetization curves (VSM graphs) for Fe_3O_4 and $Ag@Fe_3O_4$ NPs in both sulfonated C100E and SST80 polymers. (a) Comparison between Fe_3O_4 and $Ag@Fe_3O_4$ NPs in C100E polymer; (b) comparison between Fe_3O_4 and $Ag@Fe_3O_4$ NPs in C100E and SST80 polymers. (*See insert for color representation of the figure.*)

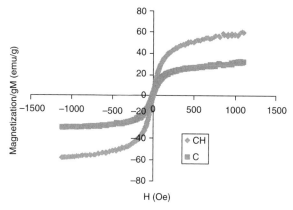

Figure 8.13. Magnetization curves (VSM graphs) for Fe_3O_4 and $Ag@Fe_3O_4$ NP magnetite prepared using lower concentrations of both metals (Fe and Ag) in sulfonic polymer, C100E (CH sample compared to C sample). (*See insert for color representation of the figure.*)

Induction, where changes in the magnetic field produce an electric field. Samples were accurately introduced in suitable test tubes and the magnetization was analyzed at room temperature.

The results of VSM measurements (Fig. 8.12 and Fig. 8.13) showed the magnetization curves for zero field cooling at room temperature of different PSMNP samples. The magnetization values were normalized according to the amount of magnetic metal according to ICP values (Table 8.3). Superparamagnetic behavior was observed in all the NCs. Figure 8.12 compares the differences in terms of magnetization among Ag@ Fe_3O_4 and Fe_3O_4 in both sulfonated polymers, C100E and SST80. Graphical representation of a material's magnetization against the strength of an applied magnetic field (H) gives rise to magnetization curve with a characteristic sigmoidal shape, where the saturation magnetization is reached if the applied magnetic field is large enough (Fig. 8.12 and Fig. 8.13).

For ferromagnetic NMs or NC materials containing ferromagnetic nanosized components, the shape of the magnetization curve often depends on the NP size. For smaller sizes (<50 nm), no hysteresis is observed due to a phenomenon called superparamagnetism. Superparamagnetism occurs in NPs that are single domain and arises as a result of magnetic anisotropy; that is, the spins are aligned along a preferred crystallographic direction. In the absence of external magnetic field, their magnetization appears to be in average zero: they are said to be in the superparamagnetic state. In this state, an external magnetic field is able to magnetize the NPs, similarly to a paramagnet. However, their magnetic susceptibility is much larger than the one of paramagnets [103].

When comparing the magnetization values of the Fe_3O_4–sulfonated polymer with Fe_3O_4 powder NPs (PF), similar magnetization values were obtained. As previously reported, Ag-coated magnetic core NPs showed a higher magnetic saturation in comparison with the uncoated magnetic NPs [104]. In all cases, $Ag@Fe_3O_4$ NCs showed higher magnetic saturation than Fe_3O_4 NCs. Therefore, it may be suggested that bimetallic Ag-coated magnetic core NPs present a synergistic effect towards magnetization as it has been already reported for Pt–Co alloy NPs [48]. Hence, in this case, the type of polymer influences the loading of the magnetic NPs, which affects the overall magnetic properties of the composite.

Figure 8.13 shows that the NCs with low amount of metals (both iron and silver elements) have higher magnetism properties in comparison with the same samples synthesized with normal concentrations (Table 8.3).

In general, the magnetic properties are mainly determined by the diameter of the crystal, its saturation magnetization, and its Néel relaxation time, which depends on the anisotropy constant [91]. The stage of aggregation of a particle as well as the particle size should also have a strong effect on the saturation magnetization because the anisotropy constant of the particles increases dramatically from the bulk value when size decreases, leading to an enhancement of the saturation [105]. Thus, the samples prepared using lower concentrations of metal are likely to have formed NPs of smaller sizes (or a lower NP aggregate dimension), leading to the observation of higher saturation magnetization. Furthermore, by a decrease of the metal concentration (for instance, sample SH, Fig. 8.13), a higher magnetization and a better distribution of the particles on the surface of the beads were obtained. Thus, we can conclude that lower concentrations during the synthesis favored the distribution of the NPs on the surface of the material (less penetrability) and enhanced the magnetic properties of the NC.

8.6 EXAMPLES AND APPLICATIONS: BIOCIDE ACTIVITY OF $Ag@Fe_3O_4$ NCs

While silver has been known to be a bactericide element for at least 1200 years, colloidal silver or silver MNPs have recently been recognized and tested in various applications as excellent antimicrobial agents because of their high biocide activity [106–108]. The antibacterial action of silver MNPs is still under debate, but it has been reported to

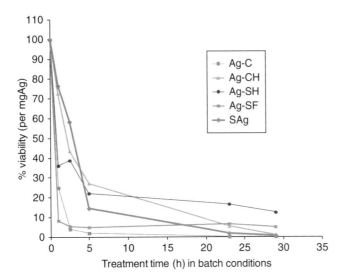

Figure 8.14. Cell viability versus treatment time for Ag@Fe$_3$O$_4$ NP NCs in both C100E and SST80 polymers and for monometallic Ag NP–SST80 NCs. (*See insert for color representation of the figure.*)

be similar to that of silver ions, which bind to DNA, block transcription, and therefore interrupt the bacterial respiration and the adenosine triphosphate synthesis. Ag MNPs purportedly present high antibacterial activity, relatively low toxicity, chemical and thermal stability, and a long-lasting action period [109]. The stabilization and immobilization of Ag MNPs in different matrices have recently gained great attention from scientists and technologists.

In the case of Ag@Co, the bactericide activity of the material as well as its lifetime was tested among different kinds of polymeric bases obtaining high activities for different kinds of bacteria [79]. Therefore, the new Ag@Fe$_3$O$_4$ NCs offer great potential and were tested and compared for antibacterial applications.

In general, the Ag NPs' antibacterial activity was evaluated by quantifying cell viability at different times after incubation with the same concentration of Ag NPs as described in previous publications [49, 79]. The antibacterial kinetics activity was determined as the relationship between the number of viable bacteria before and after the treatment in percentage terms (% viability) at several extractions/treatment times. Resin is packed in column (dimensions 3 cm long, $\emptyset_i = 0.5$ cm).

The bactericide activity of different NC structures was analyzed under batch conditions by adding 10 mL of bacterial suspensions containing 10^3 colony-forming units per mL (CFU mL^{-1}) of *Escherichia coli* to test tubes containing a known amount of NC. For the determination of the antibacterial activity, 100 μL aliquots were regularly extracted from the test tubes under sterile conditions and plated on agar containing LB medium. After overnight incubation at 37 °C, the number of viable organisms was determined by visual inspection [79]. Figure 8.14 shows the kinetics of bactericide treatment in terms of % of cell viability per mg of Ag in batch conditions for the NC samples just described (see Table 8.3).

All the Ag@Fe$_3$O$_4$ samples showed a fast decrease in cell viability initially. According to experimental data, Ag@Fe$_3$O$_4$ NC presented a decrease of more than the 90% after 2.5 h of treatment.

As shown, NCs with higher amounts of Ag metal show higher bactericide activity, as expected. In addition, NCs with higher amounts of magnetic material appear to show higher bactericide activity. This is likely to be due to the higher amount of magnetic NPs present in these samples, which provide a higher surface area for Ag coating, providing a higher active surface area of Ag for antibacterial activity. There was a significant difference in the bactericide activity when the amount of Ag was halved for both CH1 and SH1 samples. One important difference between the samples prepared with a high concentration of metal and the ones with lower concentration is the silver fractal dimension within the beads. However, the presence of fractals does not affect the bactericide activity since the bacteria cannot penetrate inside the bead, so there is no contact between bacteria and silver. The distribution of silver on the surface of the material is a crucial parameter to obtain efficient biocidal materials. Hence, these particles may not be immediately accessible to bacteria, possibly because the particles are stabilized upon entering the channels. When these particles are not accessible, bacteria use the external surface to increase the number of colonies. However, when the majority of the silver particles become accessible, the bacteria rapidly died.

To discuss these results, one has to take into account that it has been demonstrated [79] that the modified materials attack the bacteria by contact killing. Thus, a small difference in the area size of the activated surface may explain the difference in the bactericide activity kinetics. The possible reasons that may cause the surface area modification are the NP size, the NP surface, fractal presence, etc. A synergetic phenomenon by the Fe presence may be possible as it was demonstrated by the Ag@Co fibrous NCs, but confirmation of this requires further tests.

8.7 CONCLUSION

In this report we successfully described the preparation of Fe$_3$O$_4$ NPs and coated them with silver within various porous polymer matrices, of both anionic and cationic functionality. Thus, the particles with noble metal shells present a (Fe$_3$O$_4$) core–Ag shell structure by the coupling of the typical coprecipitation method for iron oxide NPs with the IMS and the Donnan exclusion effect. Characterization of the particles by XRD, TEM, SEM, TGA, and VSM or SQUID measurements authenticated the particles and coatings and their properties within the various polymer matrices. Magnetic studies reveal that the noble metal-coated particles do not change their magnetic property significantly in comparison to that of the parent Fe$_3$O$_4$ particles and this value can be enhanced for some samples. Moreover, these Ag@Fe$_3$O$_4$ particles show bactericide activity that surpasses that of the analogue Ag and Ag@Co NC within the same polymer matrix.

It was also shown that C100E and SST80 polymers (both sulfonated) gave excellent results in terms of loading and properties with magnetite (even with lower concentrations) and Ag coating.

ACKNOWLEDGMENTS

This work was supported by Research Grant MAT2006-03745, from the Ministry of Science and Technology of Spain, and by ACC1Ó for VALTEC 09-02-0057 Grant within FEDER Program. The authors also acknowledge the FI and BE (AGAUR) grants supporting A. Alonso. Special thanks are given to the Department of Genetics and Microbiology (UAB, Spain) and Centro Nacional de Microelectrònica (IMB-CNM, CSIC, Spain). Also, we acknowledge the Servei de Microscopia from Universitat Autònoma de Barcelona; the Institut de Ciències dels Materials from Barcelona (CSIC); DCU and Trinity College Dublin, from Dublin (Ireland); and the ESRF synchrotron, SpLine (Grenoble, France), for the technical service support.

REFERENCES

[1] Schmid G. *Nanoparticles: From Theory to Application.* (2nd Ed.). 2010 Wiley-VCH Verlag GmbH & Co.

[2] Ajayan et al. *Nanocomposite Science and Technology.* 2005 Wiley-VCH Verlag GmbH & Co.

[3] Carotenuto G., Nicolais L. Metal-polymer nanocomposite synthesis: novel ex-situ and in-situ approaches. In: *Metal-Polymer Nanocomposites.* G. Carotenuto, L. Nicolais (Eds.). 2005 John Wiley & Sons, Inc.

[4] Campelo J.M. et al. *ChemSusChem*, 2009, 2 (1), 18–45.

[5] Blackman J.A. *Metallic Nanoparticles.* Handbook of Metal Physics. 2009 Elsevier. pp. 385.

[6] Savage N., Diallo M. *Journal of Nanoparticle Research*, 2005, 7 (4–5), 331–342

[7] Klabunde K.J. *Nanoscale Materials in Chemistry.* 2001 John Wiley & Sons, Inc.

[8] Liu C., Zou B. Rondinone A. J., Zhang Z.J. *Journal of Physical Chemistry B*, 2000, 104, (6).

[9] Hyeon T. *Chemical Communications*, 2003, 8, 927–934.

[10] Vatta et al. *Pure and Applied Chemistry*, 2006, 78 (9), 1793–1801.

[11] Wu et al. *Nanoscale Research Letters*, 2008, 3 (11) 397–415.

[12] Klabunde K.J. *Nanoscale Materials in Chemistry.* 2005 Wiley-VCH. pp. 1–293.

[13] Macanas J. et al. Ion-exchange assisted synthesis of polymer-stabilized metal nanoparticles. In: *Solvent Extraction and Ion Exchange.* A Series of Advances, Vol. 20. 2011 Tailor & Francis.

[14] Park J., Cheon, J. *Journal of American Chemical Society*, 2001, 123, 5743–5746.

[15] Houk L.R. et al. *Langmuir*, 2009, 25, 19, 11225–1122.

[16] Imre A. et al. *Applied Physics A*, 2000, 71, 19–22.

[17] Alonso A. et al. *Environmentally-Safe Catalytically Active and Biocide Polymer-Metal Nanocomposites with Enhanced Structural Parameters.* Advances in Nanocomposite Technology. 2011. pp. 175–200.

[18] Banerjee R., Crozier P.A. *Microscopy and Microanalysis*, 2008, 14, 282–283.

[19] He J. et al. *Chemistry of Materials*, 2003, 15, 4401–4406.

[20] Porel S. et al. *Chemistry of Materials*, 2005, 17, 9–12.

[21] Buonomenna M.G. et al. *Journal of Chemical Engineering*, 2010, 5 (1), 26–34.

[22] Ruiz P. et al. *Nanoscale Research Letters*, 2011, 6, 343.

[23] Muraviev D.N. et al. *Sensors And Actuators B-Chemical*, 2006, 118 (1–2), 408–417.

[24] Auffan M. et al. *Chemical Reviews*, 2008, 108, 6, 2067.

[25] Wiesner M.R. *Nature Nanotechnology*, 2009, 4, 634–641.

[26] Haverkamp R.G. *Particulate Science And Technology*, 2010, 28 (1), 1–40.

[27] Nabok A. *Organic and Inorganic Nanostructures*. 2005 Artech House, Inc.

[28] Padovani S. et al. *Applied Physics A*, 2004, 79, 229–233.

[29] Ph. Colomban, *Journal of Nano Research*, 2009, 8, 109–132.

[30] Leonhardt U. *Nature Photonics*, 2007, 1, 207–208.

[31] Rehren T. et al. Glass coloring works within a copper-centered industrial complex in Late Bronze age Egypt. In: *The Prehistory & History of Glassmaking Technology*. P. Mc Cray (Ed.), Ceramics and Civilization Series Vol. VIII, W.D. Kingery (Series Ed.). 1998 The American Ceramic Society. p. 227.

[32] Wood N. *Chinese Glazes*. 1999 University of Pennsylvania Press.

[33] Brun N. et al. *Material Science Letters*, 1991, 10, 1418.

[34] Liem N.Q. et al. *Journal of Cultural Heritage*, 2003, 4,187.

[35] Available at http://dsc.discovery.com/news/2006/10/02/hairdye_his_zoom0.html?category= history&guid= 20061002163030 Accessed October 30, 2011

[36] Walter P. et al. *Nano Letters*, 2006, 6, 2215–2219.

[37] Ostafin A., Chen Y.C. Nanoreactors. In: *Encyclopedia of Chemical Technology* Kirk-Othmer (Ed.). 2009. 1–18.

[38] Lee H. et al. *Nano Letters*, 2004, 4 (5), 995–998.

[39] Baeuerlein, E., ed. *Biomineralization: Progress in Biology, Molecular Biology, and Application*. 2005 Wiley-VCH, 44, 4833–4834.

[40] Blakemore R.P. *Science*, 1975, 190, 377–379.

[41] Frankel R.B. *Chinese Journal of Oceanology and Limmology*, 2009, 27 (1), 1–2.

[42] Ofer S. et al. *Biophysical Journal*, 1984, 46(1), 57–64.

[43] Komeili A. et al. Proceedings of National Academy of Sciences, 2004, 101 [11], 3839–3844.

[44] Mills G.F. et al. *Industrial & Engineering Chemistry*, 1949, 41, 2842–2844.

[45] Kravchenko A.V. Kinetics and dynamics of redox sorption. In: *Ion Exchange. Highlights of Russian Science*. D. Muraviev, V. Gorshkov and A. Warshawsky (Eds.). 2000 Marcel Dekker.

[46] Kozhevnikov A.V. *Electron Ion Exchangers: A New Group of Redoxites*. 1975 Wiley.

[47] Ergozhin, E.E., Shostak, F.T. *Russian Chemical Reviews*, 1965, 34, 949–964.

[48] Alonso A. et al. *Dalton Transactions*, 2010, 39, 2579–2586.

[49] Alonso A. et al. *Chemical Communications*, 2011, 47 (37), 10464–10466.

[50] Medyak G.V. *Russian Journal of Applied Chemistry*, 2001, 74 (10), 16583.

[51] Yegiazarov Yu.G. *Reactive and Functional Polymers*, 2000, 44, 145.

[52] EPA/600/R-10/084 August 2010 www.epa.gov.

[53] Amali A. J., Rana R. K. *Green Chemistry*, 2009, 11, 1781–1786.

[54] Mandal M. et al. *Journal of Colloid and Interface Science*, 2005, 286, 187–194.

[55] Bastos-Arrieta J. et al. *Catalysis Today*, 2012, 193(1), 207–212.

[56] Corr S. A. et al. *Journal of Physical Chemistry C*, 2008, 112, 35, 13325.

[57] Schmid G. Properties. In: *Nanoparticles, from Theory to Application*. 2004 VCH.

[58] Liu G. et al. *Advanced Synthesis & Catalysis*, 2011, 353, 1317–1324.

[59] Kidambi S., Bruening M.L. *Chemistry of Materials*, 2005, 17, 301.

[60] Brock S.L., SenevRathne K. *Journal of Solid State Chemistry*, 2008, 181, 1552.

[61] Lim C.W., Lee I.S. *Nano Today*, 2010, 5 (5), 412–434.

[62] de Dios S., Diaz-Garcia E. *Analytica Chimica Acta*, 2010, 666 (1–2), 1–22.

[63] Teja A.S., Koh P.Y. *Progress in Crystal Growth and Characterization of Materials*, 2009, 55, 22.

[64] Qiao R. *Journal of Physical Chemistry C*, 2007, 111, 2426–2429.

[65] Suchorski Y. et al. *Journal of Physical Chemistry C*, 2008, 12 (50), 20012–20017.

[66] Alonso A. et al. *Catalysis Today*, 2012, 193 (1), 200–206.

[67] Gleeson O. et al. *Organic & Biomolecular Chemistry*, 2011, 9, 7929–7940.

[68] Hu A. et al. *Journal of American Chemical Society*, 2005, 127, 12486.

[69] Gleeson O. et al. *Chemistry - A European Journal*, 2009, 15, 5669.

[70] Son S.U. et al. *Journal of American Chemical Society*, 2004, 126, 5026.

[71] Guo S., Wang E. *Nano Today*, 2011, 6, 240–264.

[72] Ruiz P. et al. *Reactive and Functional Polymers*, 2011, 71, 916–924.

[73] Muraviev D.N. et al. *Pure and Applied Chemistry*, 2008, 80, 11, 2425–2437.

[74] Ruiz P. et al. *Nanoscale Research Letters*, 2011, 6, 343.

[75] Bernard, B.K. et al. *Journal of Toxicology and Environmental Health*, 1990, 29 (4), 417–429.

[76] Borm, P., Berube, D. *Nano Today*, 2008, 3 (1–2), 56–59.

[77] Chen, J.L., Fayerweather, W.E. *Journal of Occupational Medicine and Toxicology*, 1988, 30 (12), 937–942.

[78] Li et al. *Water Research*, 2008, 42 (18), 4591–4602.

[79] Alonso A. et al. *Langmuir*, 2011, 28(1), 783–790.

[80] Plietht W.J. *Journal of Physical Chemistry*, 1982, 86, 16.

[81] Davies G.L. et al. *ChemPhysChem*, 2011, 12, 772–776.

[82] Laurent S. et al. *Chemical Reviews*, 2008, 108 (6), 2064–2110.

[83] Downey D. High Total Dissolved Solids (HTDS) Produced Water Softening With PUROLITE Shallow Shell Technology Resins. AG SSTEng Report 12-12-06.

[84] Muraviev D.N. et al. *Sensors and Actuators B: Chemical*, 2006, 118 (1–2), 408–417.

[85] Muraviev D.N. et al. *Physica Status Solidi A*, 2008, 205 (6), 1460–1464.

[86] Dersch R. et al. *Polymers for Advanced Technologies*, 2005, 16, 276–282.

[87] Guyot A., Bartholin M. *Progress in Polymer Science*, 1982, 8, 277–331.

[88] Jou-Hyeon Ahn et al. *Macromolecules*, 2006, 2, 627–632.

[89] Corr S.A. et al. *Journal of Physical Chemistry C*, 2008, 112 (35), 13324–13327.

[90] Arnim Henglein, Michael Giersig, *Journal of Physical Chemistry B*, 1999, 103, 9533–9539.

[91] White R.J. et al. *Chemical Society Reviews*, 2009, 38, 481–494.

[92] Ban Z. et al. *Journal of Material Chemistry*, 2005, 15, 4660–4662.

[93] Garza-Navarro M. et al. *Journal of Solid State Chemistry*, 2010, 183, 99–104.

[94] Soomro S.S. et al. *Journal of Catalysis*, 2010, 273, 138–146.

[95] Sarkar S. et al. *Environmental Science & Technology*, 2010, 44, 1161–1166.

[96] Iglesias-Silva E. et al. *Journal of Non-Crystalline Solid*, 2007, 353, 829–831.

[97] Resina M. et al. (2007). *Journal of Membrane Science*, 289, 150–158.

[98] Fina A. et al. *Polymer Degradation and Stability*, 2006, 91.

[99] Ju Y.W. et al. *Composites Science and Technology*, 2008, 68, 1704–1709.

[100] Guerra R. et al. *Microporous and Mesoporous Materials*, 2012, 157(1), 267–273.

[101] Watanabe N. et al. *Nature*, 2005, 436, 1181–1185.

[102] Guodan Wei, Ce-Wen Nan, Yuan Deng, and Yuan-Hua Lin, *Chemistry of Materials*, 2003, 15, 4436–4441,

[103] Gittleman J.I. et al. *Physical Review B*, 1974, 9, 3891–3897.

[104] Lalatonne Y. et al. *Nature Materials*, 2004, 3, 121–125.

[105] Chen J.P. et al. *Journal of Physical Review B*, 1995, 51, 17.

[106] Kong H., Jang, J. *Langmuir*, 2008, 24 (5), 2051–2056.

[107] Law N. et al. *Applied and Environmental Microbiology*, 2008, 74 (22), 7090–7093.

[108] Pal S. et al. *Applied and Environmental Microbiology*, 2007, 73 (6), 1712–1720.

[109] Cubillo A.E. et al. *Journal of Materials Science*, 2006, 41 (16), 5208–5212.

9

PREPARATION AND CHARACTERIZATION OF ANTIMICROBIAL SILVER/ POLYSTYRENE NANOCOMPOSITES

G. Carotenuto[1], M. Palomba[1], L. Cristino[2],
M.A. Di Grazia[2], S. De Nicola[3,4], and F. Nicolais[5]

[1]*Institute for Composite and Biomedical Materials,*
National Research Council, Napoli, Italy
[2]*Istituto di Cibernetica – CNR, Pozzuoli, Napoli, Italy*
[3]*CNR-SPIN Napoli, Complesso Universitario di Monte Sant'Angelo*
via Cinthia, Napoli, Italy
[4]*INFN-Sez. di Napoli, Complesso Universitario di Monte Sant'Angelo*
via Cinthia, Napoli, Italy
[5]*Dipartimento di Scienze Politiche Sociali e della Comunicazione,*
Università degli Studi di Salerno, Fisciano, Italy

9.1 INTRODUCTION

In recent years noble metal nanoparticles have been the subjects of many researches due to their unique electronic, optical, mechanical, magnetic, and chemical properties that are significantly different from those of bulk materials [1]. These special and unique properties could be attributed to the small size and the large specific surface area characterizing these systems. For such reasons metallic nanoparticles have found applications in a variety of technological fields as catalysis, electronics, and photonics [1].

Nanocomposites: In Situ *Synthesis of Polymer-Embedded Nanostructures*, First Edition.
Edited by Luigi Nicolais and Gianfranco Carotenuto.
© 2014 John Wiley & Sons, Inc. Published 2014 by John Wiley & Sons, Inc.

Polymer-embedded metal nanoparticles represent a very simple and convenient way to use such nanostructures, and some chemical and physical techniques are now available for preparing these materials; the most important approaches are usually based on the *in situ* thermolysis of metal precursors [2].

Nanoparticles of silver have been found to exhibit interesting antibacterial activities [3, 4], and the investigation of this phenomenon has gained importance due to the increase of bacterial resistance to antibiotics, caused by their overuse. Recently, also materials (mainly textiles) containing silver nanoparticles have been developed and exhibit very interesting antimicrobial activity. Antibacterial activity of the plastic-containing silver can be used, for example, in medicine, to reduce infections as well as to prevent bacteria colonization on plastic devices like prostheses, catheters, vascular grafts, and dental materials [5]. Under proper temperature and humidity conditions, plastics can be a good medium for the generation and the propagation of microorganisms, which can cause irritations and infections. For these reasons, the polymeric materials must be protected against microorganisms in order to suppress their growth and dissemination. Owing to the high antimicrobial activity, relatively low cost, and easy production in a polymer-embedded form, nanoscopic silver could be a very adequate filler for such a purpose [6–8]. Silver nanoparticles are one of nontoxic and safe wide-range antibacterial agents to the human body, able to kill most harmful microorganisms. The mechanism of antibacterial action of silver ions is closely related to their interaction with proteins, particularly at thiol groups (sulfhydryl, -SH), and is believed to bind protein molecules together by forming bridges along them. Since the proteins behave often like as enzymes, the cellular metabolism is inhibited and the microorganism dies. Considering that the atomic radius r of the silver is much smaller than the radius R of a nanoparticle, it is possible to calculate the fraction of surface silver atoms according to the simple relationship $6r/R$. Therefore, the number of surface silver atoms increases with decreasing the dimension of the nanoparticle, and this, in turn, leads to an enhancement of the antibacterial ability of each nanoparticle [3, 4].

The purpose of this research is to study the antimicrobial characteristics of a silver/polystyrene nanocomposite system. Such advanced material has been prepared by a new chemical approach based on the thermolysis of a silver precursor dissolved in polymer. Noble metal nanoparticles can be uniformly generated into a thermoplastic polymer by controlled thermal decomposition of a metallorganic precursor dissolved in it [9, 10]. Here, silver 1,5-cyclooctadiene-hexafluoroacetylacetonate, [Ag(hfac)(COD)] (see Fig. 9.1), has been used to generate silver nanoparticles into amorphous polystyrene. This special organic compound thermally decomposes at a temperature of c. 150 °C, which is compatible with thermal stability of most thermoplastic polymers [11]. In addition, [Ag(hfac)(COD)] is characterized by an excellent solubility in nonpolar polymers and solvents, and consequently nanocomposites with a very high numerical density

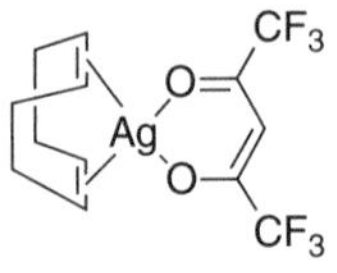

Figure 9.1. Chemical structure of the [Ag(hfac)(COD)] molecule.

of silver nanoparticles can be prepared by using such compound as silver precursor. According to the obtained results, an effective antimicrobial plastic results by thermally decomposing [Ag(hfac)(COD)] in molten polystyrene, probably because of the large amount of nanoscopic silver present on the polymer surface.

9.2 EXPERIMENTAL

9.2.1 Preparation and Characterization of the Ag/PS films

Silver 1,5-cyclooctadiene-hexafluoroacetylacetonate ($C_{13}H_{13}AgF_6O_2$, Aldrich, 99%) was purified by recrystallization from chloroform. [Ag(hfac)(COD)]/polystyrene solid solutions were prepared by dissolving the [Ag(hfac)(COD)] microcrystalline powder in chloroform and then mixing it with a viscous solution of amorphous polystyrene (Aldrich, $M_w = 230,000$ gmol^{-1}) in chloroform. In order to investigate the dependence of nanocomposite antimicrobial properties on the nanoparticle concentration, different [Ag(hfac)(COD)]/polystyrene samples with compositions ranging from 1% to 30% by weight of [Ag(hfac)(COD)] were prepared. The high solubility of silver 1,5-cyclooc tadiene-hexafluoroacetylacetonate in amorphous polystyrene at room temperature allows to prepare nanocomposite systems with quite large amounts of silver nanoparticles without formation of silver aggregates. The obtained liquid solutions were cast onto a Petri dish and allowed to dry in air at 25 °C for 48 h. Then, nanocomposite films were produced by isothermally heating the dry [Ag(hfac)(COD)]/polystyrene systems onto a laboratory hot plate at a temperature of 180 °C. The precursor [Ag(hfac)(COD)]/ polystyrene films were transparent and colorless, but during the controlled thermal annealing treatment, a yellow coloration of the films was developed that evolved into brown by increasing the precursor concentration (a silvery aspect was observed at higher concentrations).

9.2.2 Bacterial Culture and Viability Assays

The original culture of *Escherichia coli* was made in sterile lysogeny broth (LB) (Invitrogen Life Corporation, California, United States) in sterile tubes at starting dilution of 10^{-2} and incubated over 24 h at 37 °C under gentle agitation. Therefore, 1 ml of this starting culture was transferred in 9 ml of fresh broth in triplicate sampling, for each of the treatments summarized in Table 9.1 (A–F), over 16, 24, and 48 h incubation at 37 °C under gentle agitation.

By spectrophotometric analysis the increased turbidity of the reading reflects the index of bacterial growth and cell numbers (biomass), and the amount of transmitted light decreases as the cell population increases. The absorbance, or optical density (OD), was read at wavelength 540 nm, and it gives an indirect measurement of the number of *E. coli*. To this purpose, the absorbance of 1 ml of each specimen was measured by the spectrophotometer (Bio-Rad, Denmark). The spectrophotometric measurement was made by standardizing the machine on the sterile nutrient broth, being its concentration equal to zero.

TABLE 9.1. Time-Dependent Antiseptic Evaluation of Ag/PS Films. Spectrophotometric Lecture of the OD in *E. coli* Culture Significant ($*P \leq 0.05$) and Very Significant ($**P \leq 0.001$) Antibacterial Effect After 24 h and 48 h of Incubation with Ag/PS Films, Both at Mild and Drastic Annealing

Samples (*E. coli* culture)	16 h OD	24 h OD	48 h OD
(A). Blank, without Ag/PS film	0.855	1.003	1.205
(B). +10wt.% Ag/PS film 1 cm × 1 cm 0,0150 mg	0.774	0.898	0.939
(C). +30wt.% Ag/PS film 1 cm × 1 cm 0,0177 mg **drastic annealing**	0.687	0.815($*P \leq 0.05$)	($**P \leq 0.001$)
(D). +30wt.% Ag/PS film 1 cm × 1 cm 0,0177 mg **Mild annealing**	0.786	0.822($*P \leq 0.05$)	0.862 ($*P \leq 0.05$)
(E). +0.009 mg pure Ag powder	0.591 ($*P \leq 0.05$)	0.722 ($*P \leq 0.05$)	0.891 ($*P \leq 0.05$)
(F). +0.007 mg pure Ag powder	0.656 ($*P \leq 0.05$)	0.890	1.056

The spectrophotometric measurements were performed at 16, 24, and 48 h of incubation with different composition of Ag/PS films, and they were repeated for each treatment (A–F in Table 9.1).

This spectrophotometric analysis was paralleled by the LIVE/DEAD *Bac*Light Bacterial Viability assay for microscopy (Invitrogen Life Corporation, California, United States), which allows fast and high-sensitive, qualitative and quantitative, monitoring of bacterial viability as a function of the membrane integrity of the cell. This method is based on two nucleic acid stains: green-fluorescent SYTO® 9 stain and red-fluorescent propidium iodide stain, which differ in their ability to penetrate healthy bacterial cells. SYTO® 9 stain labels live bacteria; in contrast, propidium iodide penetrates only bacteria with damaged membranes. Quantitative analysis was performed by cell counting in Bürker chamber on live/dead stained bacterial culture after 16, 24, and 48 h of Ag/PS films incubation compared to blank and Ag pure powder treatments. The results were expressed as percentage of live or dead bacteria calculated on the total number of cells.

9.3 RESULTS AND DISCUSSIONS

The minimum temperature value required for [Ag(hfac)(COD)] thermal decomposition was established by differential scanning calorimetry (DSC, TA Instrument Q100) and thermogravimetric analysis (TGA, TA Instruments Q500) performed on the pure compound. As visible in Figure 9.2, this organic salt melts at 110 °C, and according to the TGA test, it simultaneously decomposes, losing the 1,5-cyclooocatadiene (COD) ligand. Then, the decomposition of the Ag(hfac) fragment follows; this process is strongly exothermal and starts at ca. 150 °C. The exothermal peak is deformed by the simultaneous evaporation of the thermolysis by-products, which is an endothermal phase transition. In fact, according to the TGA performed in air at 10 °min⁻¹ (see Fig. 9.3),

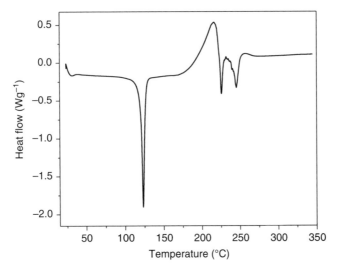

Figure 9.2. Differential scanning calorimetry thermogram of a pure [Ag(hfac)(COD)] sample.

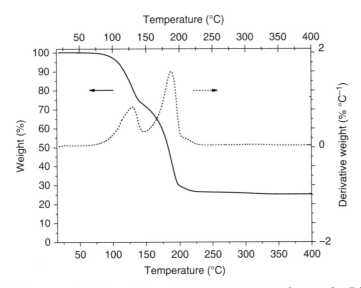

Figure 9.3. Thermogravimetric analysis and DTGA thermograms of a pure [Ag(hfac)(COD)] sample.

weight loss of organic by-product is visible at 150 °C, and a residual weight of 25%, which exactly corresponds to the formation of zerovalent silver, resulted.

The formation of a zerovalent silver nanoscopic phase by the thermal decomposition of [Ag(hfac)(COD)] dissolved in amorphous polystyrene was confirmed by large-angle X-ray powder diffraction (XRD) analysis. In particular, the main peak in the XRD diffractogram shown in Figure 9.4 corresponds to (111) signal of the ccp-silver

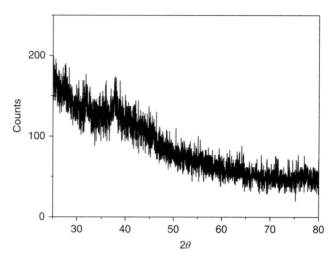

Figure 9.4. X-ray powder diffraction of Ag/PS nanocomposite obtained by thermally decomposing [Ag(hfac)(COD)] dissolved in polystyrene.

diffraction pattern. Owing to the nanoscopic silver size, a quite low signal-to-noise ratio characterizes the XRD diffractogram. Therefore, pure [Ag(hfac)(COD)] was quantitatively converted to zerovalent silver by the performed thermal treatment.

The TEM and SEM micrographs shown in Figure 9.5 represent the typical microstructure of a Ag/PS nanocomposite sample. In particular, the sample was obtained by annealing an [Ag(hfac)(COD)]/polystyrene film (30% by weight of [Ag(hfac)(COD)]) for 10 s at 180 °C. Pseudo-spherical and quite monodispersed silver nanoparticles, of c. 7 nm, appear to be homogeneously distributed into the continuous polystyrene matrix. In addition, a large amount of very small silver nanoparticles (c. 3 nm) were present everywhere over the sample. According to the SEM micrographs, silver nanoparticles were present also at the surface of the silver/polystyrene nanocomposite.

Differently from the use of thiolates as thermolytic precursor for the *in situ* generation of metal nanoparticles into an amorphous polymer matrix [2], the silver 1,5-cyclooctadiene-hexafluoroacetylacetonate precursor originates nanoparticles with a larger average particle size. Probably, such a behavior is related to the presence of a capping agent (i.e., hexafluoroacetylacetonate molecules), less effective than the sulfur-containing molecules (e.g., disulfide) present in the thiolate precursor thermal decomposition.

A UV–Vis spectrophotometer (PerkinElmer, Mod. LAMBDA 850) was employed to study the surface plasmon resonance of the polymer-embedded silver nanoparticles obtained at low concentration of precursor. Figure 9.6 shows the extinction spectrum of a [Ag(hfac)(COD)]/polystyrene film (30% by weight of [Ag(hfac)(COD)]). As visible, an approximately Gaussian-shaped peak characterized the surface plasmon absorption of the silver nanoparticle dispersion into the dielectric polystyrene matrix, which is indicative of a quite non-aggregated dispersion of nanoparticles for different precursors' content (thermal annealing of 10 s at 180 °C).

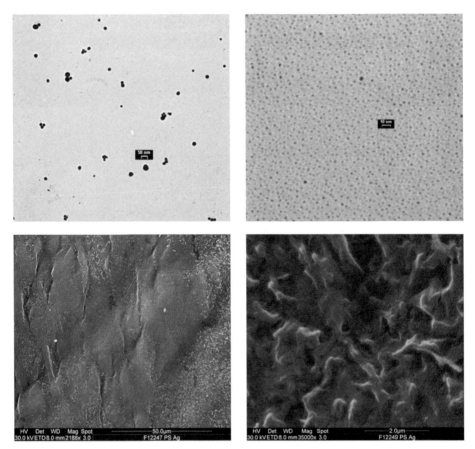

Figure 9.5. TEM and SEM micrographs of a nanocomposite sample. As visible, the silver particles are present also at film surface.

The antibacterial activity of a small piece of silver/polystyrene film is summarized in Table 9.1. Ag nanoparticles are present at film surface, and therefore they can release silver ions in the environment where they are placed, while [Ag(hfac)(COD)] has been completely converted into gaseous products (COD, CO_2, H_2O, etc.) and silver. In order to investigate the dependence of the antibacterial activity on the silver nanoparticle concentration, two different [Ag(hfac)(COD)]/PS] nanocomposites containing 30% by weight of precursor have been tested: the first is obtained by using a mild thermal annealing treatment (20 s at 150 °C) and the second is obtained by a drastic thermal annealing treatment (60 s at 150 °C).

The antiseptic properties were evaluated by spectrophotometric lecture of the OD measured at standard wavelength (540 nm), using a fixed volume of *E. coli* suspension culture, continuously stirred up to 16, 24, and 48 h of incubation. Table 9.1 reports the OD values of the bacteria incubated with different Ag/PS nanocomposite samples at different times (16, 24, and 48 h), the OD value of a blank sample (i.e., bacteria culture

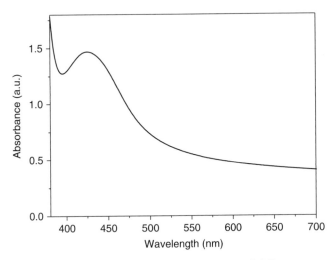

<u>Figure 9.6.</u> Optical spectrum of [Ag(hfac)(COD)]/PS film of different compositions.

without the Ag/PS film), and the values of bacteria culture incubated with pure metallic silver powder at different concentrations.

From data reported in Table 9.1, it is visible as the OD of samples incubated with plastic films is less than that of the blank sample and only slightly less than that of the same sample incubated with two different concentrations of pure silver. This behavior confirmed the good antibacterial activity of such nanocomposite material. It can be also seen that the OD of the blend films decreased with the increasing of the silver content, thus indicating an antibacterial activity increase with the Ag nanoparticles content.

Sample containing the same percentage by weight of silver precursor (30wt.%) but obtained after a drastic annealing treatment showed higher antibacterial properties compared to that obtained after a mild annealing treatment. Furthermore, the statistical analysis of the data at 24 h and 48 h of incubation showed a value of statistic significance $P \leq 0.05$ with OD values of the sample of Ag/PS (30% weight of precursor, mild annealing) and the sample incubated with 0.009 mg of pure Ag powder and $P \leq 0.001$ with the OD value of the sample of Ag/PS (30% weight of precursor, drastic annealing). Such parameter is provided to confirm that the obtained data are statistically significant ($P \leq 0.05$).

Furthermore, in order to investigate on the existence of a faster Ag/PS antibacterial action, we performed qualitative and qualitative evaluation of Ag/PS activity for each of the treatments (A–E in Table 9.1) after 16 h incubation.

To this purpose we exploited the high-sensitive qualitative and quantitative LIVE/DEAD® BacLight™ Bacterial Viability assay for fluorescence microscope [12]. This method is based on the combined ability of the membrane-permeant SYTO® 9 to label live bacteria with green fluorescence and membrane-impermeant propidium iodide to label membrane-compromised bacteria with red fluorescence. This procedure was performed for each Ag/PS treatment, and the antimicrobial effects were analyzed by Leica DMI6000 epifluorescence microscope using appropriate emission filters.

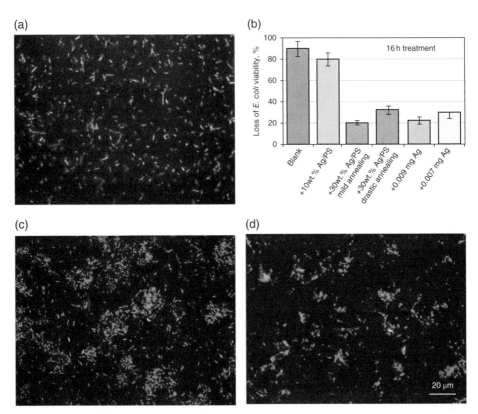

Figure 9.7. Live/dead cell staining assay: microscopic analysis and quantitative evaluation. (a) High number of green-labeled living bacteria after 16 h incubation in LB, without Ag/PS. (c, d) High number of red-labeled dead bacteria after 16 h incubation +30 wt.% Ag/PS film (1 cm × 1 cm − 0.0177 mg) obtained by using a mild thermal annealing treatment (C); similar results were observed in bacteria culture after 16 h of incubation with Ag powder (D). Scale bar represents 20 μm. (b) Quantitative analysis of loss of *E. coli* viability (%) after 16 h incubation at each condition reported in Table 9.1 (A–E). (*See insert for color representation of the figure.*)

The images were acquired using a digital camera (Leica DFC420), and data analysis was performed using the Leica MM AF Analysis Offline software (Leica, Germany). Following the procedure described in Reference [13], the quantitative analysis of loss cell viability was calculated by the ratio of dead on live cells counted in the Bürker chamber. The data were expressed as mean of percentage ± s.d.m., and a repeated two-way ANOVA was performed using SPSS version 8.2. Student's or Bonferroni two-tail t-test was used to judge statistical significance between groups ($P < 0.05$).

The image shown in Figure 9.7a was recorded after 16 h of incubation in the presence of the liquid medium for bacteria growth (LB). It clearly displays a persistent green fluorescence extended to the entire bacteria culture. However, the system becomes significantly modified in the case of bacteria culture incubated with plastic film of Ag/PS (30% by weight of precursor, mild thermal annealing).

Indeed, Figure 9.7c, recorded after 16 h of incubation, shows that almost all the bacteria are red colored with a very little number of green-colored cells. These qualitative effects were confirmed by quantitative analysis of antimicrobial properties. The corresponding results, summarized in Figure 9.7b, indicate an intense loss of % *E. coli* viability after mild and drastic thermal annealing with 30% by weight of precursor.

Similar effect was found in pure Ag powder, which is in good agreement with the data obtained by the spectrophotometric OD technique. However, the sensitivity of this method is lower, and it shows a significant antimicrobial effect only after 24 h of ageing time of the Ag/PS sample.

The antibacterial activity of a zerovalent silver phase is strictly depending on the surface development of the solid since silver atoms/ions required to accomplish the antibacterial activity are released only from the surface. Consequently, when this solid phase is in a powdered form, the resulting antibacterial activity can be significantly increased, and an ultrafine silver powder may become several orders of magnitude more active than the corresponding bulk solid. Therefore, the incorporation of nano-sized silver (silver clusters or nanoparticles) into a plastic material should represent the best choice to fabricate materials devoted to such a technological application.

To investigate the influence on the antibacterial activity of a silver atom distribution, a simple geometrical model based on a silver particle of radius R and an atom of radius $r \ll R$ can be considered. Then, it is possible to calculate the number of surface silver atoms according to the simple relation: $4R^2/r^2$, which corresponds to the fraction $6r/R$ (percentage of surface silver atoms). These simple considerations show that the fraction of surface silver atoms increases with decreasing of size of the particle and that we may assume that all atoms in a nano-sized powder are practically on the surface. All atoms in a hyperfine powder are, to a good approximation, available for the antibacterial activity because they can leave the particle and move to the sulfured groups (i.e., sulf-hydryls, –SH, and disulfides, –S–S–) present in the proteins of the microbe cellular membrane.

Actually the silver atoms are arranged in a cubic crystal, and a more detailed analysis should also take into account the coordination numbers of the atoms in the crystal edges and corners and those located at the basal planes. The coordination numbers are 3, 4, and 5 for the atoms in the crystal edge, corner, and basal plane, respectively. If we assume a cubic crystal of dimensions L and a lattice constant much less than L, simple dimensional considerations show that the fraction of atoms in the crystal edge and corners increases with decreasing of the size L of the particle compared to the fraction of atoms at the basal planes. This means that the atoms in the crystal edge and corner can leave the crystal surface more easily and diffuse in the surrounding liquid medium where microbes and other microorganisms are contained and that this tendency is enhanced in a nano-size particle compared to a larger one.

Finally, a nano-sized silver antibacterial activity is enhanced for two main reasons: (i) the fraction of surface silver atoms increases with decreasing the particle size, and

(ii) there is greater fraction of surface silver atoms only weakly bonded to the particle surface that can be easily released in the surrounding medium.

9.4 THE ANTISEPTIC MECHANISM OF SILVER NANOPARTICLES

Antiseptic mechanism for silver ions is widely discussed in a review paper of Rai et al. [14]. Silver ions enter in the bacterial cells by diffusing through the cell wall and turn the DNA into a condensed form, which reacts with thiol groups contained in the proteins, thus causing cell death. Furthermore, silver ions also interfere with the replication process. The mechanism of action of silver is based on the interaction of silver ions with thiol groups in the respiratory enzymes of bacterial cells. Silver binds to the bacterial cell wall and to cell membrane and inhibits the respiration process. In case of *E. coli*, silver acts by inhibiting the uptake of phosphate and releasing phosphate, mannitol, succinate, proline, and glutamine from *E. coli* cells. However, although the mechanism for the antimicrobial action of silver ions is not completely understood, the effect of silver ions on bacteria can be recovered by the morphological changes induced in the cells. It is suggested that when DNA molecules are in relaxed state, the replication of DNA can be effectively conducted. On the other hand, when DNA is in condensed form, it loses its replication ability. Therefore, when the silver ions penetrate inside the bacterial cells, the DNA molecule turns into condensed form and loses its replication ability, thus leading to cell death.

Silver nanoparticles show efficient antimicrobial property compared to other salts due to their extremely large surface area, which provides better contact with microorganisms. The nanoparticles get attached to the cell membrane and also penetrate inside the bacteria. The bacterial membrane contains sulfur-containing proteins, and the silver nanoparticles interact with these proteins in the cell as well as with the phosphorus-containing compounds like DNA. When silver nanoparticles enter the bacterial cell, it forms a low molecular weight region in the center of the bacteria to which the bacteria conglomerates, thus protecting the DNA from the silver ions. The nanoparticles preferably attack the respiratory chain, which in turn results in an inhibition of cell division and leads to cell death. Furthermore, the nanoparticles release silver ions in the bacterial cells, thus enhancing their bactericidal activity.

The bactericidal effect of silver nanoparticles is size and shape dependent. The size of the nanoparticle implies that it has a large surface area to come in contact with the bacterial cells. Particles smaller than 10 nm have higher percentage of interaction than bigger ones. The surface plasmon resonance, which plays an important role in the determination of optical absorption spectra of metal nanoparticles, shows a sensible dependence on the particle size, since it shifts to a longer wavelength with increase in particle size.

The shape dependence of the antimicrobial efficacy of the nanoparticle has been confirmed by studying the inhibition of bacterial growth by differentially shaped nanoparticles. Truncated triangular nanoparticles exhibit antimicrobial activity with less silver content compared to the case of spherical silver particles and to rod-shaped

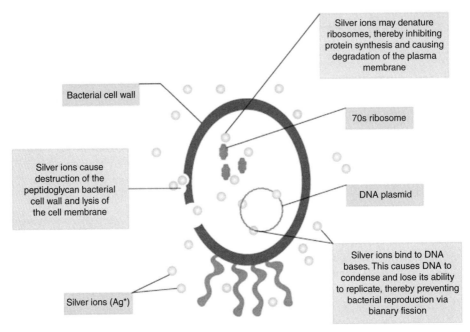

Figure 9.8. Mechanisms of the antibacterial activity of silver ions. It is widely accepted that the major antibacterial effect of silver nanoparticles is mediated by its partial oxidation and release of silver ions and. Silver ions interact with the peptidoglycan cell wall and the plasma membrane, causing cell lysis; bacterial (cytoplasmic) DNA, preventing DNA replication; and bacterial proteins, disrupting protein synthesis. Multifaceted antibacterial activity is the key to low bacterial resistance rates observed for silver and silver nanoparticles that also can directly damage and penetrate the cell wall and plasma membrane.

particles. An explicative Figure 9.8 that well summarizes the antiseptic mechanism explained is reported in the following text.

9.5 CONCLUSION

Antimicrobial silver/polystyrene nanocomposites have been prepared by dissolving silver 1,5-cyclooctadiene-hexafluoroacetylacetonate into amorphous polystyrene and thermally annealing this solid solution for 10 s at 180 °C. The silver atoms generated during the [Ag(hfac)(COD)] thermal decomposition clusterize, leading to the formation of a hyperfine dispersion of silver nanoparticles. The obtained silver-doped polystyrene material shows very strong antimicrobial characteristics, which are comparable with the antimicrobial activity of a pure micrometric silver powder for silver concentrations in the polymer higher than 30% by weight.

REFERENCES

[1] Nicolais L., Carotenuto G., "Metal-polymer nanocomposites" *John Wiley & Sons Incorporation.: Hoboken, New Jersey*, 2005.

[2] Nicolais F., Carotenuto G., "Synthesis of polymer-embedded metal, semimetal, or sulfide clusters by thermolysis of mercaptide molecules dissolved in polymers", *Recent Patent on Materials Science, 1* (2008) 1–11.

[3] Amato E., Diaz-Fernandez Y.A., Taglietti A., Pallavicini P., Pasotti L., Cucca L., Milanese C., Grisoli P., Dacarro G., Fernandez-Hechavarria J.M., Necchi V., "Synthesis, Characterization and Antibacterial Activity against Gram Positive and Gram Negative Bacteria of Biomimetically Coated Silver Nanoparticles" *Publications Organization of American Chemical Society/Langmuir*, 2011, 27, 9165–9173.

[4] Pallavicini P., Taglietti A., Dacarro G., Diaz-Fernandez Y.A., Galli M., Grisoli P., Patrini M., Santucci De Magistris G., Zanoni R., "Self-assembled monolayers of silver nanoparticles firmly grafted on glass surfaces: Low Ag+ release for an efficient antibacterial activity", *Journal of Colloid and Interface Science*, 2010, *350*, 110–116.

[5] Dallas P., Sharma V.K., Zboril R., "Silver polymeric nanocomposites as advanced antimicrobial agents: Classification, synthetic paths, applications, and perspectives", *Advances in Colloid and Interface Science*, 2011, *166*, 119–135.

[6] Deshmukh R.D., Composto R.J., "Surface segregation of silver nanoparticles in the in-situ synthesized Ag/PMMA nanocomposites", *Bulletin of the American Physical Society*, 2006 March Meeting.

[7] Damm C., Munstedt H., "Kinetic aspects of the silver ion release from antimicrobial polyamide/silver nanocomposite", *Applied Physics A*, 2008, *91*, 479–486.

[8] Jokar M., Russly A. R., Nor Azowa I., Lugman Chuah A., Tan C.P., "Melt Production and Antimicrobial Efficiency of Low-Density Polyethylene (LDPE)-Silver Nanocomposite Film", *Food Bioprocess Technology*, 2012, *5*, 719–728.

[9] Carotenuto G., Palomba M., Longo A., De Nicola S., Nicolais L., "Optical limiters based on silver nanoparticles embedded in amorphous polystyrene", *Science and Engineering of Composites Materials*, 2011, *18*, 187–190.

[10] Pullini D., Carotenuto G., Palomba M., Mosca A., Horsewell H., Nicolais L., "In situ synthesis of high-density contact-free Ag-nanoparticles for plasmon resonance polystyrene nanocomposites", *Journal of Materials Science*, 2011, *46*, 7905–7911.

[11] Willis A.L., Chen Z., He J., Zhu Y., Turro N.J., O'Brien S., "Metal acetylacetonates as general precursors for the synthesis of early transition metal oxide nanomaterials", *Journal of Nanomaterials*, 2007, *2007*, 1–7.

[12] Berney M., Hammes F., Bosshard F., Weilenmann H.U., Egli T., "Assessment and Interpretation of Bacterial Viability by Using the LIVE/DEAD BacLight Kit in Combination with Flow Cytometry", *Applied and Environmental Microbiology*, 2007, *73*, 3283–3290.

[13] Sjollema J., Rustema-Abbing M., van der Mei H.C., Busscher H.J., "Generalized Relationship between numbers of bacteria and their viability in biofilms", *Applied and Environmental Microbiology*, 2011, 77, 5027–5029.

[14] M. Rai, A. Yadav, A. Gade, "Silver nanoparticles as a new generation of antimicrobials", *Biotechnology Advances* 27(2008)76–83.

10

NANOMATERIAL CHARACTERIZATION BY X-RAY SCATTERING TECHNIQUES

C. Giannini, D. Siliqi, and D. Altamura

Institute of Crystallography, CNR, Bari, Italy

10.1 WHY X-RAYS?

In general, a source of rays, suitable optics, and a detector are required to observe an object. Atoms, however, are far too small to be discerned using any visible light source because atomic radii range from a few tenths of an angstrom to a few angstroms, and they are smaller than 1/1000 of the wavelengths present in visible light (from ~4000 to ~7000 Å). X-rays, which are shortwave electromagnetic radiation discovered by W. C. Roentgen, have a suitable wavelength to observe individual atoms (1). Indeed they have the wavelengths that are commensurate with both the atomic sizes and short-est interatomic distances in crystalline matter. Unfortunately, as the index of refraction of X-rays is ~1 for all materials, they could not be focused by a lens until very recently (2). This explained why X-ray microscopes appear much later than visible light micro-scopes or electron microscopes. In crystallography, X-rays are not used to image individual atoms directly. However, as was first shown by Max von Laue in 1912 using a single crystal of hydrated copper sulfate ($CuSO_4.5H2O$), the periodicity of the crystal lattice allows atoms in a crystal to be observed with exceptionally high resolution and precision by means of X-ray diffraction. As we will see later, the diffraction pattern of a crystal is a transformation of an ordered atomic structure into the reciprocal space (Fourier space) rather than a direct image of the crystal. The three-dimensional

Nanocomposites: In Situ *Synthesis of Polymer-Embedded Nanostructures*, First Edition.
Edited by Luigi Nicolais and Gianfranco Carotenuto.
© 2014 John Wiley & Sons, Inc. Published 2014 by John Wiley & Sons, Inc.

distribution of atoms in a lattice can be restored only after the diffraction pattern has been transformed back into direct space, by means of robust crystallographic methods of inversion. The process of transforming diffraction patterns in order to reinstate the underlying crystal structures in the three-dimensional direct space is governed by the theory of diffraction. The latter rests on several basic assumptions, yet it is accurate and practical. We have no intent to cover the comprehensive derivation of the X-ray diffraction theory since it is mainly of interest to experts and can be found in many excellent books and review [3]. In this chapter we will discuss the nature and sources of X-ray-based techniques that are in common use today to characterize materials, especially nanomaterials.

10.2 NANOMATERIALS AND X-RAYS

Nowadays, nanomaterials can be fabricated through top-down and bottom-up approaches. Top-down refers to the situation where the nanosized objects are constructed by breaking up larger entities, for instance, with the motivation to improve the applicability of certain properties. In the bottom-up approach, the nano-objects are built from smaller entities, in principle atom by atom. The nanosystems built in this way show properties without counterparts in the macroscopic world, and their nanoscopic size enables novel applications that have been inaccessible so far. Fundamental is to understand and characterize the very smallest structural details of these nano-objects in order to explain, improve, and exploit its properties.

Many different characterization techniques have been developed over the years, especially the different types of scanning probe microscopy as well as (high-resolution) transmission electron microscopy ((HR)TEM), and have proven most useful, not only for characterization purposes but also to device design. These microscopy techniques are able to give detailed information on the structure of a limited number of nano-objects; thus, they are excellent local probes. As opposed to this, a number of more global probes exist: extended X-ray absorption fine structure (EXAFS), X-ray absorption near-edge structure (XANES), small-angle X-ray scattering (SAXS), X-ray powder diffraction (XRPD), or wide-angle X-ray scattering (WAXS) are some of the most important ones. Generally speaking, these techniques give information on the average structure in a sample containing a large number of nanoparticles. The experiments still render information on the atomic arrangement (including defects) and shape and size of the particles, but the interpretation is complicated by the inevitable presence of impurities and defects. The complementarity of the different characterization techniques is often needed to get a differentiated understanding of the material at hand.

In this chapter we will introduce basic concepts on SAXS and WAXS, with their counterpart for grazing geometries, namely, grazing incidence SAXS and WAXS (GISAXS and GIWAXS) and XRPD.

Depending on the specific technique, the diffraction setup will be always composed by (i) an X-ray source (either a Roentgen tube, a rotating anode, a microsource, a synchrotron source, or a free-electron laser), (ii) primary optics (pinhole, monochromator, grating, mirror, Fresnel zone plate, waveguide, etc.), (iii) the sample attachment to mount the sample either in a capillary holder or onto a goniometer in flat plate, (iii) secondary optics (slits, analyzer, mirror, etc.), and (iv) a 0D, 1D, or 2D detector. The technical choices will depend on the chemical/physical nature of the typical materials to be analyzed with that instrumentation (organic, inorganic, powder, nanomaterials, thin film, etc.). The X-ray pattern is collected in the reciprocal space (Fourier space), and, according to the chosen geometry and type of measurements, it allows one to probe different regions of the reciprocal space and consequently to provide structural and morphological information at different length scales (from the atomic to the mesoscopic scale).

10.3 WIDE- AND SMALL-ANGLE X-RAY SCATTERING (WAXS AND SAXS)

A simple and intuitive description of the X-ray diffraction phenomena was given by Bragg (4). When a beam of parallel and monochromatic X-rays impinges onto a crystal whose lattice planes are displaced by an interplanar distance d_{hkl}, the diffracted beam forms an angle θ with the crystal planes that can be derived by the relationship $2d_{hkl}\sin\theta = n\lambda$, known as the Bragg law.

The Bragg law can be written also as

$$\left|\vec{r}^*\right| = \frac{2\sin\theta}{\lambda} = \frac{1}{d_{hkl}}$$

where $\vec{r}^*$ is the lattice in the reciprocal space.

In any case, given $\vec{k}$ and $\vec{k}'$, the incident and scattered momentum, the momentum transferred that describes the elastic scattering is given by

$$\vec{q} = \vec{k} - \vec{k}', \left|\vec{q}\right| = \frac{2\pi}{\lambda}2\sin\theta = \frac{2\pi}{d}$$

Here $\left|\vec{k}\right| = \left|\vec{k}'\right| = \dfrac{2\pi}{\lambda}$ where λ is the wavelength of the incoming beam (see Fig. 10.1).

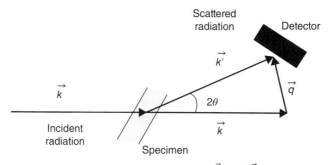

Figure 10.1. Typical scheme of a scattering experiment: $\vec{k}$ and $\vec{k}'$, the incident and scattered momentum.

(a)   (b)   (c)

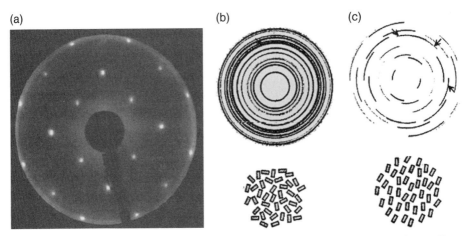

Figure 10.2. Diffraction pattern of (a) a single crystal, (b) random assembly of crystallites (polycrystal), (c) polycrystal with preferred orientations. Reprinted from http://www.crystal. mat.ethz.ch/research/ZeolitesPowderDiffraction/Texture. (*See insert for color representation of the figure.*)

The transferred vector gives the length scale that is accessible through a specific scattering process.

All structures are investigated in a noninvasive and nondestructive way.

We will now describe two types of scattering process, called WAXS and SAXS.

In WAXS, wide is the angle accessed during measurements, typically varying from 5° to 100°. As a consequence, data contain information of the sample at the length scales of the atoms (atomic resolution). Indeed, these data are the results of the interference process due to the secondary waves emitted by the electronic density in the lattice. From the registered angular distribution of this pattern, in principle, it is possible to perform a full structural characterization of the investigated specimen reaching atomic resolution.

The diffraction pattern, especially if collected with a bidimensional detector, allows to make a clear distinction on the type of order in the investigated specimen that can be (i) a single crystal made with a perfect 3D crystalline order across mm^2 (Fig. 10.2a), (ii) a polycrystal made of several randomly ordered crystallites (Fig. 10.2b), and (iii) a polycrystal made of preferred-oriented crystallites (Fig. 10.2c).

When the 2D pattern is azimuthally integrated, the 1D-folded profile (Fig. 10.3) contains diffraction peaks of different intensities and widths at half maximum. Depending on the type of pattern, several information can be extracted [5]:

- Quantitative phase analysis (QPA): qualitative and quantitative analysis of the crystalline phases present in the specimen
- Fraction of crystalline/amorphous phases
- Microstructural analysis: dimension and orientation of the crystalline domains and relative microstrain

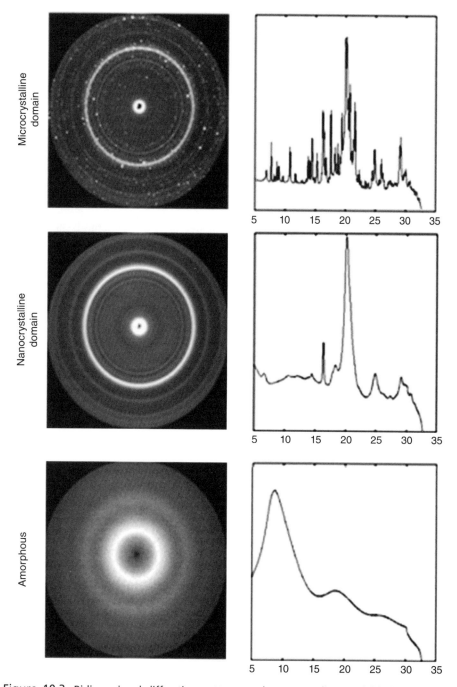

Figure 10.3. Bidimensional diffraction patterns and corresponding 1D-folded azimuthally integrated profiles for micro- and nanocrystal and for amorphous. Reprinted from Reference [6], with permission from ELSEVIER.

- Structural analysis: type, positions, and symmetry relation of the atoms in the unit cell, unit cell size, and space group (unit cell content)

Figure 10.3 shows three different diffraction profiles collected on specimens containing crystallites with microcrystalline- or nanocrystalline-sized domains or no crystallites at all (amorphous) [6].

For diffraction patterns measured from polycrystalline samples, the Rietveld method is one of the most widespread approaches to extract the listed information. The method was first introduced by Hugo Rietveld [7] to refine data from neutron scattering and was later expanded for the use of X-ray data analysis. A general guideline for newcomers in the field of Rietveld refinement can be found in the literature [8] and a more detailed description in Young [9]. Typically a Rietveld refinement can be performed if the structure (unit cell) is approximately known. The strategy of Rietveld refinement is to simulate a physical model, including structural, sample, and instrumental parameters and fit it with the least square method to the experimental observed data [10].

Whenever a crystallographic model of the structure does not exist, the Rietveld approach cannot be applied. This happens, for example, for same nanocrystals as Au and Ag noble metals, which, below about 10 nm, typically form crystallographically forbidden phases, such as decahedron and icosahedron [11]. In these cases, a Debye function (DF)-based full-pattern fitting method has been proposed for powder diffraction data treatment [11–13].

Especially in the field of nanocrystal characterization, the microstructural analysis is extremely important and can be achieved in two ways:

1. An estimate of the domain size can be achieved through the Scherrer formula [14]:

$$D = \frac{\lambda \kappa}{\beta_{h,k,l} \cos \theta}$$

   where κ is a constant, λ is the X-ray wavelength, θ is the Bragg angle, and $\beta_{h,k,l}$ is the integral breadth of that specific diffraction peak. This approach is a bit more accurate when the instrumental resolution function contribution is negligible (small nanocrystals smaller than 10 nm).
2. A whole profile fitting analysis of the pattern using the Rietveld or the DF approach (preferable choice), modeling a shape anisotropy.

As an example, Figure 10.4 displays the diffraction pattern collected from CdSe nanocrystals having the same crystal structure (wurtzite) but different domain shape (sphere for the black profile and rod for the red one). Indeed, the (002) diffraction peak becomes sharper due to the elongation of the nanoparticles along the [001] direction. The shape anisotropy affects the entire pattern, and therefore any whole profile fitting approach is better suitable to extract it from data with higher accuracy than the simple Scherrer formula.

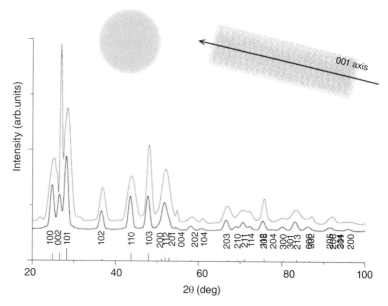

Figure 10.4. Diffraction profiles of CdSe wurtzite nanocrystals with different shapes: sphere (black profile) and rod (red profile). (*See insert for color representation of the figure.*)

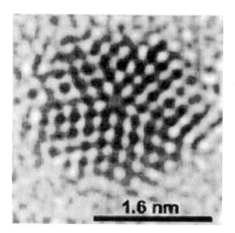

Figure 10.5. Multiple-domain gold nanocrystals, as inspected by HRTEM. Reprinted with permission from Reference [17]. Copyright 1997 American Chemical Society.

In SAXS, the signal arises from electron density inhomogeneities and allows to obtain structural information of the probed material with characteristic length scales in the order of 10 to 1000 Å^{-1} [15, 16]. Since those distances are large compared to those of electrons or even atoms in the material, it is not possible to separate their contribution to the scattering. The summation over the individual electrons is therefore replaced by the integral over the average electron density.

The scattered radiation is registered in an angular range that is extremely small (below $2°$). As an example, with a $\lambda = 1.54 \text{Å}$ wavelength and θ varying between $0.0°$ and $1.3°$, it is possible to have access to information across length scales between 34Å and 1470Å. Each experimental setup has been thought for different needs and will allow to achieve slightly different length scales.

At the nanoscale (1–500 nm), matter is characterized by basic complex elements, which are seldom periodic. Indeed, the SAXS signal carries on morphological information of any nano-object that, independently of its crystallinity, is embedded into a matrix. In other words, the nano-object can be studied with SAXS provided that its visibility (contrast in the electronic density) allows it to be detected. It is worth to address two main differences between SAXS and WAXS:

1. Small-angle X-ray scattering does not require any nano-object crystallinity, while the WAXS signal cannot be registered in absence of lattice periodicity.
2. Wide-angle X-ray scattering provides information on the crystalline domain, while SAXS on the nano-object external shape. These are often different as in the case of the gold nanoparticles [15] shown in Figure 10.5.

SAXS could never be sensitive to the crystalline domain borders as the electron density does not change between them.

The scattered intensity I(q) of not interacting particles with well-defined shaped particles embedded in a medium of homogeneous electron density can be expressed analytically by using a form factor P(q,R). Form factors for various shapes were reported by Pedersen [18]. The shape of the particles can be extracted from I(q) but only if they are monodisperse or exhibit a very narrow size distribution. However, this is not the case for many materials, and therefore the effect of polydispersity has to be taken into account by a particle size distribution. The effect of the size distribution on I(q) is illustrated in Figure 10.6.

So far I(q) was described neglecting any effect of particle–particle interactions, but this is only the case for particles in very diluted systems. In dense materials, I(q) might be influenced by interference due to particle interactions. The effect of the particle–particle interactions on I(q) can be described by a structure factor S(q,R). For monodisperse, randomly oriented, and centrosymmetric particles, I(q) can be described by $I(q) = (\Delta\rho_r)^2 P(q,R)^2 \, S(q,R)$. Different models for structure factors can be found in the literature [18, 19] and will not be discussed in detail here.

At sufficient low scattering vectors, $qR < 1$, the scattered intensity I(q) can be approximated by Guinier's law

$$I(q) \cong \left(\Delta\rho_r^2 V^2\right) \exp\left(-\frac{q^2 R_g^2}{3}\right)$$

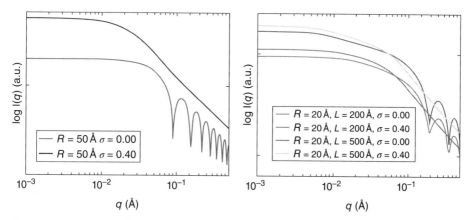

Figure 10.6. Calculated scattering pattern of mono- and polydisperse spheres with a (mean) radius of 50 Å (left) and of mono- and polydisperse cylinders with different lengths (right). The scattered intensities for polydisperse particles were simulated with a lognormal size distribution.

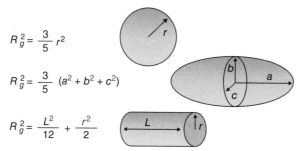

$$R_g^2 = \frac{3}{5} r^2$$

$$R_g^2 = \frac{3}{5} (a^2 + b^2 + c^2)$$

$$R_g^2 = \frac{L^2}{12} + \frac{r^2}{2}$$

Figure 10.7. Radius of gyration (R_g) for different shapes. (*See insert for color representation of the figure.*)

where R_g is the radius of gyration and is defined as the mean square from the center of gravity of the particle, in terms of electron density, in analogy to the radius of gyration in classical mechanics. The radius of gyration for common shaped particles studied with small-angle scattering are shown in Figure 10.7 and, can be found in the literature [16].

Changing the shape of the nano-object, the SAXS profile changes accordingly [17], as represented in Figure 10.8.

For diluted specimens (negligible structure factor), the Guinier analysis can be easily performed plotting $\ln(I(q))$ versus q^2. The analysis allows to extract the radius of gyration and to guess the nano-object shape. Fitting of the SAXS profile can be performed, starting from this preliminary hypothesis, refining the size and shape of the scattering object.

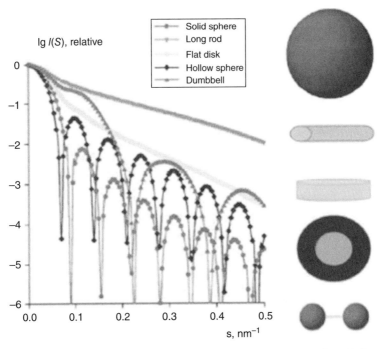

Figure 10.8. SAXS patterns for different shapes. Reprinted with permission from Reference [20]. *(See insert for color representation of the figure.)*

10.4 GRAZING INCIDENCE WIDE- AND SMALL-ANGLE X-RAY SCATTERING (GIWAXS AND GISAXS)

The theoretical basis of the GIWAXS and GISAXS techniques is partially the same as for WAXS and SAXS. A fundamental difference distinguishes them: GIWAXS/GISAXS is used to study materials lying onto the top of surfaces or underneath them, especially to characterize layered devices; WAXS/SAXS is used with transmission geometries for materials in form of powders, solutions, or nanoparticles embedded into a host matrix. Indeed, in GISAXS the presence of surface/interfaces makes the scattering process more complex than for SAXS geometry. Once all terms involved in the scattering are correctly taken into account [21], GISAXS, as such, can provide information on the particle form factor (morphological information) or on the nanoscale periodic structure, typically occurring in the case of assembled nanocrystals. GIWAXS, instead, contains information on the atomic structure of the investigated samples, which is studied at grazing incidence angle. In both GISAXS and GIWAXS, the incidence angle can be varied, changing the penetration depth of the X-ray beam and consequently probing different sample thicknesses below the surface. With the possibility to acquire at the same instrument, without moving the sample, both GIWAXS and GISAXS can offer the interesting possibility to have both information (morphological/structural at the nanoscale and atomic) at the same time.

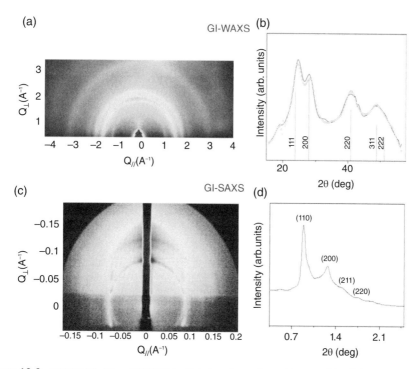

Figure 10.9. GIWAXS (**a, b**) and GISAXS (**c, d**) data collected on a 3D assembly of PbS nanocrystals. Reprinted with permission from Reference [25]. (*See insert for color representation of the figure.*)

As an example, Figure 10.9 shows the GIWAXS (a, b) and GISAXS (c, d) data collected [24] on a 3D assembly of PbS nanocrystals [22–24], as deposited onto a Si substrate. The azimuthally integrated diffraction rings in Figure 10.9b match with the diffraction profile of rock-salt (CsSl-type) PbS nanocrystals; the full width at half maximum of the peak can be justified by about 2 nm domain size [25]. The corresponding GISAXS image and corresponding 1D-folded profile, shown in Figure 10.9c and Figure 10.9d, was indexed as a bcc superlattice [26].

In general, two different types of GISAXS/GIWAXS experiments can be realized: *ex situ* experiments focused on static samples, prepared *ex situ* and investigated decoupled from preparation, which can be performed with laboratory equipment [25, 27], and real-time, *in situ* measurements [28, 29] typically performed with synchrotron radiation. Among them, more recent developments were realized combining small X-ray beams obtained by micro-focusing and nano-focusing optics, scanning sample position, and GISAXS to enable a position-dependent GISAXS information. These so-called micro-beam and nano-beam GISAXS experiments were introduced and realized at dedicated beamlines [30–34].

Today key features of GISAXS/GIWAXS are the ability to access:

- Buried structures, which are located well below the surface and thus are inaccessible to local probe techniques such as atomic force microscopy or scanning electron microscopy

- A depth-dependent structural information using different incident angles
- Structures in any kind of environments ranging from ultrahigh vacuum to gas atmospheres and liquids (at synchrotron radiation facility)
- Kinetic studies performed as function of external parameters such as temperature, gas pressure, pH, or ion concentration

10.5 X-RAY POWDER DIFFRACTION (XRPD)

X-ray powder diffraction is one of the mostly used nondestructive tools to analyze matter, especially for powders and polycrystalline materials. From research to production and engineering, XRPD is an indispensible method for structural material characterization and fast quality control. This technique uses X-ray diffraction on powders or polycrystalline samples, where ideally every possible crystalline orientation is represented equally. The resulting orientational averaging causes the three-dimensional reciprocal space that is typically studied in single crystal diffraction to be projected onto a single dimension (see Fig. 10.2). This collapse of information causes severe peak overlapping that makes the diffraction pattern analysis more difficult. X-ray powder diffraction patterns can be acquired both in transmission geometry (Debye–Scherrer) and in reflection geometry (Bragg–Brentano). To eliminate possible effects of texturing and to achieve true randomness, the sample is generally rotated during Debye–Scherrer measurements. The diffractogram is like a unique "fingerprint" of the investigated material. The XRPD method gives laboratories the ability to quickly analyze unknown materials and characterize them in such fields as metallurgy, mineralogy, forensic science, archeology, and the biological and pharmaceutical sciences. Identification is performed by comparison of the diffractogram to known standards or to international databases. The information contained in a diffraction pattern is:

1. Peak positions encoding information on crystal system, space group symmetry, translation symmetry, and unit cell dimensions
2. Peak intensities encoding information on unit cell contents and point symmetry
3. Peak shapes and widths encoding information on crystallite size, nonuniform microstrain, and extended defects (stacking faults, antiphase boundaries, etc.)

As for WAXS, XRPD allows to provide valid solutions to the following list of items:

- Qualitative analysis: phase identification and crystalline phases in a mixture
- Quantitative phase analysis: lattice parameter determination and crystalline phase fraction analysis
- Structure solution and refinement through Rietveld method, ab initio reciprocal space methods, and ab initio real space methods
- Peak shape analysis: crystallite size distribution, microstrain analysis, and extended defect concentration

REFERENCES

[1] Landwehr, G. (1997). Hasse, A. ed. Röntgen centennial: X-rays in Natural and Life Sciences. Singapore: World Scientific. 7–8.

[2] Di Fabrizio, E., Romanato, F., Gentili, M., Cabrini, S., Kaulich, B., Susini, J. and Barrett, R. (1999). *Nature* **401**, 895–898.

[3] Giacovazzo C., Monaco H. L., Artioli G., Viterbo D., Milanesio M., Ferraris G., Gilli G., Gilli P., Zanotti G., Catti M. (2011). Fundamentals of Crystallography (3rd edition), Oxford University Press.

[4] Bragg, W. L. (1913). Proceedings of the Cambridge Phylosophical Society **17**: 43–57.

[5] Dorfs, D., Krahne, R., Giannini, C., Falqui, A., Zanchet, D. and Manna L. (2011). Quantum Dots: Synthesis and Characterization, from Comprehensive Nanoscience and Technology (vol. 2) Elsevier, Editors: Guozhong Cao, Duncan Gregory, Thomas Nann.

[6] Guagliardi, A, Cedola, A., Giannini, C., Ladisa, M., Cervellino, A., Sorrentino, A., Lagomarsino, S., Cancedda R. and Mastrogiacomo M. (2010). *Biomaterials* **31**, 8289–8298.

[7] Rietveld, H. M. (1969). *J. Appl. Crystallogr.* **2**, 65–71.

[8] McCusker, L. B., von Dreele, R. B., Cox, D. E., Louër, D., Scardi, P. (1999). *J. Appl. Crystallogr.* **32**, 336–50.

[9] Young, R. A. (1993) The Rietveld Method, Oxford: Oxford University Press

[10] FullProf http://www.ill.eu/sites/fullprof/.

[11] Cervellino, A., Giannini, C., Guagliardi, A. (2003). *J. Appl. Crystallogr.* **36**, 1148–1158.

[12] Cervellino, A., Giannini, C., Guagliardi, A. and Ladisa M. (2005). *Phys. Rev. B* **72**, 035412.

[13] Cervellino, A., Giannini, C., Guagliardi, A. (2010). *J. Appl. Cryst.* **43**, 1543–1547.

[14] Patterson, A. (1939). *Phys. Rev.* **56** (10) 978–982.

[15] Glatter O., Kratky, O. (1982). Small Angle X-ray Scattering, New York: Academic Press.

[16] Feigin, L. A. and Svergun, D. I. (1987). Structure Analysis by Small-Angle X-ray and Neutron Scattering, New York: Plenum Press.

[17] Schaaff, T. G., Shafigullin, M. N., Khoury, J. T., Vezmar, I., Whetten, R. L., Cullen, W. G. and First P. N. (1997). *J. Phys. Chem. B* **101**, 7885–7891.

[18] Pedersen J. S. (1997). *Adv. Colloid Interface Sci.* **70** 171–210.

[19] Kohlbrecher, J. (2010). SASfit: A program for fitting simple structural models to small angle scattering data, Villigen: PSI.

[20] Svergun, D. I & Koch, M. H. J. (2003). *Rep. Prog. Phys.* **66**, 1735–1782.

[21] Renaud, G., Lazzari, R., Leroy, F. (2009) *Surf. Sci. Rep.* **64**, 255–380

[22] Altamura, D., Corricelli, M., De Caro, L., Guagliardi, A., Falqui, A., Genovese, A., Nikulin, A. Y., Curri, M. L., Striccoli, M. and Giannini C. (2010). *Cryst. Growth Des.* **10**, 3770–3774.

[23] Corricelli, M., Altamura, D., De Caro, L., Guagliardi, A., Falqui, A., Genovese, A., Agostiano, A., Giannini, C., Striccoli M. and Curri M. L. (2011). *Cryst Eng Comm.* **13**, 3988–3997.

[24] Altamura, D., De Caro, L., Corricelli, M., Falqui, A., Striccoli, M., Curri M. L. and Giannini C., (2012). *Cryst. Growth Des.* **12**, 1970–1976.

[25] Altamura, D., Lassandro, R., Vittoria, F. A., De Caro, L., Siliqi, D., Ladisa M. and Giannini C., (2012). *J. Appl. Crystallogr.* 45, 869–873.

[26] Goodfellow, B. W., Patel, R. N., Panthani, M. G., Smilgies, D.-M.and Korgel, B. A. (2011). *J. Phys. Chem. C* **115**, 6397–6404.

[27] Chiu, M.-Y., Jeng, U.-S., Su, C.-H., Liang, K. S., and Wei, K.-H., (2008). *Adv. Mater.* **20**, 2573–2578.

[28] Doshi, D. A., Gibaud, A., Liu, N., Sturmayr, D., Malanoski, A. P., Dunphy, D. R., Chen, H., Narayanan, S., MacPhee, A., Wang, J., Reed, S. T., Hurd, A. J., van Swol, F., Brinker, C. (2003). *J. Phys. Chem. B* **107**, 7683 62.

[29] Lee B., Yoon J., Oh W., Hwang Y., Heo K., Jin K. S., Kim J., Kim K. W., Ree M. (2005). *Macromolecules* **38**, 3395.

[30] Roth S. V., Burghammer M., Riekel C., Muller-Buschbaum P., Diethert A., Panagiotou P., (2003). *Appl. Phys. Lett.* **82**, 1935. 63, 69, 78.

[31] Roth S. V., Rankl M., Artus G. R. J., Seeger S., Burghammer M., Riekel C., Muller-Buschbaum P (2005). *Physica B* **357**, 190.

[32] Muller-Buschbaum P., Roth S. V., Burghammer M., Diethert A., Panagiotou P., Riekel C. (2003). *Europhys. 3 Lett.* **61**, 639.

[33] Muller-Buschbaum P., Roth S. V., Burghammer M., Bauer E., Pfister S., David C., Riekel C. (2005). *Physica B* **357**, 148.

[34] Muller-Buschbaum P., Perlich J., Abul Kashem M.M., Schulz L., Roth S. V., Cheng Y. J., Gutmann (2007). *Phys. Status Solidi* (RRL) **1**, R119.

INDEX

Note: Page numbers in *italics* refer to Figures; those in **bold** to Tables.

A520E polymer, 179, **180, 182**
Ag(hfac)(COD) *see* silver 1,5-cyclooctadiene-hexafluoroacetylacetonate (Ag(hfac) (COD))
Ag@Co and Pd@Co NC synthesis
  Ag-520E NC, 175, *175*
  Co NP distribution, 173, *174*
  core-shell NPs preparation, 175
  DEDIMS protocol, 173
  edge energy, Ag and Co standards, 176, *176*
  merge spectrum, Co, 176, *176*
  metal reduction conditions, 175
  oxidation, 177
  PSMNPs preparation, 172–3
  stabilization, carboxylated fibrous materials, 173, *174*
  XANES technique, 175
Ag(hfac)-(1,5-COD), 28, 34
Ag-PDMS *see* silver-poly(dimethylsiloxane) (Ag-PDMS)
antimicrobial silver/polystyrene nanocomposite
  Ag/PS films, 197
  antibacterial activity, silver/polystyrene film, **198,** 201
  antiseptic mechanism, silver nanoparticles, 205–6, *206*
  bacterial culture and viability assays, 197–8, **198**
  differential scanning calorimetry thermogram, 198, *199*

live/dead cell staining assay, 203, *203*
optical spectrum, [Ag(hfac)(COD)]/PS film, *202*
thermogravimetric analysis and DTGA thermograms, 198, *199*
UV-vis spectrophotometer, 200
XRD, 199–200, *200*
zerovalent silver nanoscopic phase formation, 199
antiseptic mechanism, silver nanoparticles, 205–6, *206*
Au-PDMS nanocomposites *see* gold-poly(dimethylsiloxane) (Au-PDMS) nanocomposites

biocide activity, Ag@Fe$_3$O$_4$ NCs, 188–90
Brownian tree-type fractals, 186

chemical vapor deposition (CVD), 19
cobalt nanoparticles
  9.9% and 21.9% w/w cobalt, *54,* 54–5
  dissolution, [Co$_2$(CO)$_8$], 52
  infrared spectra, filtered and dried particles, 53, *53*
  magnetic measurements, polychloroprene-cobalt nanocomposites, 54, *54*
  oxidation, 52
  re-dispersion, 53
  size and shape, 52
  synthesis procedure, 52, 53, *53*
CVD *see* chemical vapor deposition (CVD)

Nanocomposites: In Situ *Synthesis of Polymer-Embedded Nanostructures*, First Edition.
Edited by Luigi Nicolais and Gianfranco Carotenuto.
© 2014 John Wiley & Sons, Inc. Published 2014 by John Wiley & Sons, Inc.